BeagleBone Home Automation Blueprints

Automate and control your home using the power of the BeagleBone Black with practical home automation projects

Rodolfo Giometti

[PACKT] open source ✱
PUBLISHING   community experience distilled

BIRMINGHAM - MUMBAI

BeagleBone Home Automation Blueprints

Copyright © 2016 Packt Publishing

First published: January 2016

Production reference: 1280116

Published by Packt Publishing Ltd.
Livery Place
35 Livery Street
Birmingham B3 2PB, UK.

ISBN 978-1-78398-602-6

www.packtpub.com

Credits

About the Author

Rodolfo Giometti is an engineer, IT specialist, and GNU/Linux expert. He has over 15 years of experience working with platforms based on x86 (PC), ARM, MIPS, and PowerPC. He is the first author and a maintainer of the LinuxPPS project (Linux's pulse per second subsystem.) He actively contributes to the Linux source code with several patches and new device drivers for several devices. His areas of core expertise are in writing device drivers for new peripherals, realizing special Linux ports for several custom embedded boards, and developing control automation and remote monitoring systems. He is the cofounder of the Cosino Project, where he develops software for industry control automation systems. He is also the co-CEO at HCE Engineering S.r.l., a leading industrial embedded systems-based company.

I would like to thank my wife Valentina and my children Romina and Raffaele for their patience during the writing of this book. I also give deep thanks and gratitude to the guys at Packt: Vivek Anantharaman, who gave to me the opportunity to write this book, and Rashmi Suvarna, who supported me in finishing this book.

Many thanks to Armando Genovese for his help in designing the circuitries, to Michele Scafoglieri and Elango Palanisamy for their support and effort in reviewing this book so carefully and to Vatsal Surti for his help in fixing up my English.

In the end, I cannot forget to thank my parents who, by giving me my first computer when I was a child, allowed me to do what I do today.

About the Reviewers

Armando Genovese is an electronics engineer with over 25 years of experience in analog and digital design, management, and marketing. Specializing in embedded circuits and systems, he has carried out projects for various industrial, medical, and consumer applications. Genovese is the founder of HCE Engineering S.r.l., which specializes in the creation of embedded systems, and is one of the founders of the Cosino Project. Genovese received a master's degree in electronics from Università degli Studi di Pisa.

Elango Palanisamy has a bachelor's degree in electronics and communication engineering from Anna University, Chennai. He is currently pursuing a master's in embedded system technologies at the same university. He has experience in firmware, board bring-ups, power management in consumer electronic devices, Linux driver development, and optimizations for thermal printers and car multimedia-related sensors. He also has experience of using build tools such as Yocto Project and Buildroot for Ti, Freescale, and Atmel boards. He currently works for HCL technologies, Chennai.

I would like to thank my parents and friends for their help and support throughout my career.

Michele Scafoglieri is an electronic engineer who designs software systems. In recent years, he has worked as a project manager for a software house and as a control engineer for manufacturing and logistics companies. Michele has over 20 years of experience in project management, software development, and embedded systems. He handled numerous projects ranging from small embedded devices to automated warehouses. Nowadays, as an independent consultant, he likes to speak at seminars, write articles, and use his skills to help others build useful systems for healthcare, logistics, and many other fields.

He can be reached at http://www.studiosigma.pro.

www.PacktPub.com

Support files, eBooks, discount offers, and more

For support files and downloads related to your book, please visit www.PacktPub.com.

Did you know that Packt offers eBook versions of every book published, with PDF and ePub files available? You can upgrade to the eBook version at www.PacktPub.com and as a print book customer, you are entitled to a discount on the eBook copy. Get in touch with us at service@packtpub.com for more details.

At www.PacktPub.com, you can also read a collection of free technical articles, sign up for a range of free newsletters, and receive exclusive discounts and offers on Packt books and eBooks.

https://www2.packtpub.com/books/subscription/packtlib

Do you need instant solutions to your IT questions? PacktLib is Packt's online digital book library. Here, you can search, access, and read Packt's entire library of books.

Why subscribe?

- Fully searchable across every book published by Packt
- Copy and paste, print, and bookmark content
- On demand and accessible via a web browser

Free access for Packt account holders

If you have an account with Packt at www.PacktPub.com, you can use this to access PacktLib today and view 9 entirely free books. Simply use your login credentials for immediate access.

Table of Contents

Preface

The BeagleBone Black is an embedded system that is able to run an embedded GNU/Linux distribution as well as normal (and powerful) distributions like Debian or Ubuntu, and to which the user can connect several external peripherals via two dedicated expansion connectors.

Because it has powerful distribution capabilities with an easily expandable embedded board, the BeagleBone Black system is a state-of-the-art device that allows the user to build powerful and versatile monitoring and controlling applications.

This book presents several home automation prototypes in both hardware and software in order to explain to you how we can use the BeagleBone Black board connected with several devices to control your home.

Each prototype is discussed in its respective chapter in both hardware and software, explaining all the connections and the software packages necessary to manage several peripherals. Then, the code to glue all of it together is explained in detail till the final test of each project.

The hardware devices used in this book have been chosen in order to cover all the possible connection types we can encounter while working with a BeagleBone Black board, so you will find I²C, SPI, USB, 1-Wire, serial, and digital and analog devices.

The programming languages used in this book have been chosen according to the rule to find the quickest and easiest solution to solve the current problem; in particular, you can find example code in Bash, C, PHP, Python, HTML, and even JavaScript.

Warning! All the projects in this book are prototypes and cannot be used as final applications.

Neither the author of this book nor Packt Publishing recommend or endorse these products to be used alone or as a component in any application without the necessary modifications to turn these prototypes into final products.

Neither the author of this book nor Packt Publishing will be held liable for unauthorized use of these prototypes. The user can use both the hardware and software of these devices at their own risk!

In the chapters where we will need to use a daemon or a kernel module, or where we will need to recompile the whole kernel, I've added a short description about what the reader should do and where they can get more information regarding the tools used; however, some basic skills in managing a GNU/Linux system, the kernel's modules, or the kernel itself are required (the reader can take a look at the book *BeagleBone Essentials*, *Packt Publishing*, written by the author of this book, in order to have more information about these topics).

What this book covers

Chapter 1, Dangerous Gas Sensors, will show how to use the BeagleBone Black to monitor some dangerous gases such as CO, Methane, LPG, and so on in a room and how to enable an acoustic and visive alarm in case of danger. Also, by using a GSM module, the user will be able to send an SMS message to a predefined phone number to alert, for instance, a relative.

Chapter 2, Ultrasonic Parking Assistant, will show how to use the BeagleBone Black to implement a park assistant. We will use an ultrasonic sensor to detect the distance between our car and the garage wall, and some LEDs to give feedback to the driver about the car's position in order to avoid collisions.

Chapter 3, Aquarium Monitor, will show how to make an aquarium monitor through which we'll be able to record all the environment data as well as control the life of our beloved fish from a web panel on our PC, smartphone, or tablet.

Chapter 4, Google Docs Weather Station, will take a look at a simple weather station that can also be used as an Internet-of-Things device. This time, our BeagleBone Black will collect environment data and will send them to a remote database (a Google Docs spreadsheet) in order to be reworked and presented into a shared environment.

Chapter 5, WhatsApp Laundry Room Monitor, will present an implementation of a laundry monitor room with several sensors capable of alerting the user directly on their WhatsApp account when a specific event occurs.

Chapter 6, Baby Room Sentinel, will present a possible implementation of a baby room sentinel capable of monitoring the room by detecting if the baby is crying or if the baby is actually breathing during sleep. Also, as a special feature, the system will be able to measure the baby's temperature with a contactless temperature sensor.

Chapter 7, Facebook Plant Monitor, will show how to implement a plant monitor capable of measuring light, soil moisture, and soil temperature (in the soil and outside it), and how to take some pictures, at specific intervals, via a webcam, and then publishing them on a Facebook timeline.

Chapter 8, Intrusion Detection System, will show how we can implement a low cost intrusion detection system with a reasonable quality level by using our BeagleBone Black and two (or more) webcams. The system will be able to alert the user by sending an e-mail message with a photo of the intruder.

Chapter 9, Twitter Access Control System with Smart Card and RFID, will show how to use a smart card reader as well as two kinds of RFID reader (LF and UHF) in order to show different ways to implement a minimal identification system for access control that can send an alert message to a Twitter account.

Chapter 10, A Lights Manager with a TV Remote Controller, will show how to manage a simple on/off device connected to our BeagleBone Black by using a TV remote controller or any infraredcapable device.

Chapter 11, A Wireless Home Controller with Z-Wave, will present how to implement a little wireless home controller by using a Z-Wave controller connected to our BeagleBone Black and two Z-Wave devices: a wall plug and a multisensor device.

What you need for this book

You need the following prerequisites to get the most from this book.

Software prerequisites

Regarding the software, the reader should have a little knowledge of a non-graphical text editor such as vi, emacs, or nano. Even if the reader can connect an LCD display, a keyboard, and a mouse directly to the BeagleBone Black and can use the graphical interface, we assume in this book that the reader is able to do little modifications to text files by using a text-only editor.

The host computer, that is, the computer the reader will use to cross-compile the code and/or manage the BeagleBone Black, is assumed to be running a GNU/Linux-based distribution. My host PC is running an Ubuntu 14.04 LTS, but the reader can also use a Debian-based system with a little modification, or they may use another GNU/Linux distribution, but with a little effort mainly for the installation of the cross-compiling tool. Systems such as Windows, Mac OS, or similar are *not* covered in this book due to the fact that we should not use low technology systems to develop code for a high technology system!

Knowing how a C compiler works and how to manage a `Makefile` could help, but don't worry, all examples start from the very beginning so that even a non-skilled developer should be able to do the job.

This book will present some kernel programming techniques, but it must not be taken as a *kernel programming course*. You need a proper book for such a topic! However, each example is well documented, and the reader will find several suggested resources.

Regarding the kernel, I'd like to state that by default I used the onboard kernel, that is, version 3.8.13. However, in some chapters, I used a self-compiled kernel, version 3.13.11; in that case, I gave a little tutorial about how to do the job.

If you're using a newer kernel release you may get minor issues, but you should be able to port whatever I did without problems. In the case that you are using a very recent kernel, please consider that the cape manager file /sys/devices/bone_ capemgr.9/slots has been relocated to /sys/devices/platform/bone_capemgr/ slots, so you have to change all related commands accordingly.

As a final note, I assume that the reader knows how to connect a BeagleBone Black board to the Internet in order to download a package or a generic file.

Hardware prerequisites

In this book, all code is developed for the BeagleBone Black board revision C, but the reader can use an older revision without any issues; in fact, the code is portable and it should work on other systems too!

Regarding the computer peripherals used in this book, I reported in each chapter where I got the hardware and where the reader can buy it, but, of course, they can decide to surf the Internet in order to find a better and cheaper offer. A note on where to find the datasheet is also present.

The reader should not have any difficulty in connecting the hardware presented in this book with the BeagleBone Black since the connections are very few and well documented. Readers don't require any particular hardware skills (apart from knowing how to use a soldering iron); however, a small amount of knowledge of electronics may help.

During the reading, I'm going to refer to BeagleBone Black's pins, especially the ones on the expansion connectors. All the pins used are explained, but, if needed, you can find a complete description of BeagleBone Black's connectors at `http://elinux. org/Beagleboard:Cape_Expansion_Headers`.

Who this book is for

If you are a developer who wants to learn how to use embedded machine learning capabilities and get access to a GNU/Linux device driver to collect data from a peripheral or to control a device, this is the book for you.

If you wish to manage your home by implementing different kinds of home automation devices that can interact with a smartphone, tablet, or PC, or if you are just working standalone, have some hardware or electrical engineering experience then this book is for you. Knowing the basics of C, Bash, Python, PHP, and JavaScript programming in a UNIX environment is also required.

Conventions

In this book, you will find a number of text styles that distinguish between different kinds of information. Here are some examples of these styles and an explanation of their meaning.

Codes and command lines

Code words in text, database table names, folder names, filenames, file extensions, pathnames, dummy URLs, user input, and Twitter handles are shown as follows: "To get the preceding kernel messages, we can use both the `dmesg` and `tail -f /var/log/kern.log` commands."

A block of code is set as follows:

```
CREATE TABLE status (
    n VARCHAR(64) NOT NULL,
    v VARCHAR(64) NOT NULL,
    PRIMARY KEY (n)
) ENGINE=MEMORY;
```

Any command line input or output given on the BeagleBone Black is written as follows:

```
root@beaglebone:~# make CFLAGS="-Wall -O2" helloworldcc -Wall -O2
helloworld.c -o helloworld
```

Any command line input or output given on my host computer as a *non-privileged user* is written as follows:

```
$ tail -f /var/log/kern.log
```

When I need to give a command as a privileged user (root) on my host computer the command line input or output is then written as follows:

```
# /etc/init.d/apache2 restart
```

The reader should notice that all the privileged commands can also be executed by a normal user by using the `sudo` command with the form:

```
$ sudo <command>
```

So, the preceding command can be executed by a normal user as:

```
$ sudo /etc/init.d/apache2 restart
```

Kernel and logging messages

On several GNU/Linux distributions, a kernel message has the following form:

```
Oct 27 10:41:56 hulk kernel: [46692.664196] usb 2-1.1: new high-speed USB
device number 12 using ehci-pci
```

This is a quite long line for this book, which is why the characters from the beginning of the line until the point where the real information begins are dropped. So, in the preceding example, the line's output will be reported as follows:

```
usb 2-1.1: new high-speed USB device number 12 using ehci-pci
```

Long outputs and repeated or less important lines in a terminal are dropped by replacing them with three dots (...), as follows:

```
output begin
output line 1
output line 2
...
output line 10
output end
```

File modifications

When the reader should modify a text file, I'm going to use the unified context `diff` format, since this is a very efficient and compact way to represent a text modification. This format can be obtained by using the `diff` command with the -u option argument.

As a simple example, let's consider the following text in `file1.old`:

```
This is first line
This is the second line
This is the third line
. . .
. . .
This is the last line
```

Suppose we have to modify the third line, as highlighted in the following snippet:

```
This is first line
This is the second line
This is the new third line modified by me
. . .
. . .
This is the last line
```

The reader can easily understand that reporting the whole file each time for a simple modification is quite obscure and space consuming; however, by using the unified context `diff` format, the preceding modification can be written as follows:

```
$ diff -u file1.old file1.new
--- file1.old 2015-03-23 14:49:04.354377460 +0100
+++ file1.new 2015-03-23 14:51:57.450373836 +0100
@@ -1,6 +1,6 @@
 This is first line
 This is the second line
-This is the third line
+This is the new third line modified by me
 . . .
 . . .
 This is the last line
```

Now, the modification is very clear and is written in a compact form! It starts with a two-line header where the original file is preceded by - - - and the new file is preceded by +++, then there are one or more changed hunks that contain the line differences in the file. The preceding example has just one hunk where the unchanged lines are preceded by a space character, while the lines to be added are preceded by a + character, and the lines to be removed are preceded by a – character.

Serial and network connections

In this book, I'm going to mainly use two different kinds of connections to interact with the BeagleBone Black board: the serial console and an SSH terminal. The former can be accessed directly via the connector J1 (never used in this book) or via an emulated way over the same USB connection that is used to power up the board, while the latter can be used via the the USB connection above or via an Ethernet connection.

The serial console is mainly used to manage the system from the command line. It's largely used to monitor the system, especially to take the kernel messages under control.

An SSH terminal is quite similar to the serial console even if is not exactly the same (for example, kernel messages do not automatically appear on a terminal); however, it can be used in the same manner as a serial console to give commands and edit files from the command line.

In the following chapters, I'm going to use a terminal on the serial console or over an SSH connection indifferently to give most of the commands and configuration settings needed to implement all the prototypes explained in this book.

To get access to the USB emulated serial console from your host PC, you can use the minicon command as follows:

```
$ minicom -o -D /dev/ttyACM0
```

Note that on some systems, you may need root privileges to get access to the /dev/ttyACM0 device (in this case, you can use the sudo command to override it).

As stated above, to get access to the SSH terminal, you can use the emulated Ethernet connection over the same USB cable that was used for the serial console. In fact, if your host PC is well configured, when you plug in the USB cable to power up your BeagleBone Black board, after a while, you should see a new cable connection with the IP address 192.168.7.1. Then, you can use this new connection to get access to your BeagleBone Black by using the following command:

```
$ ssh root@192.168.7.2
```

The last available communication channel is the Ethernet connection. It is used mainly to download files from the host PC or the Internet, and it can be established by connecting an Ethernet cable to the BeagleBone Black's Ethernet port and then configuring the port according to your LAN settings.

However, it's quite important to point out that you can also get connected to the Internet by using the emulated Ethernet connection that was presented before. In fact, by using the following commands on the host PC (obviously GNU/Linux based), you'll be able to use it as a router, allowing your BeagleBone Black board to surf the Internet as if it was connected to its real Ethernet port:

```
# iptables --table nat --append POSTROUTING --out- interface eth1 -j
MASQUERADE
# iptables --append FORWARD --in-interface eth4 -j ACCEPT
# echo 1 >> /proc/sys/net/ipv4/ip_forward
```

Then, on the BeagleBone Black, we should set the gateway through the USB cable by using the following command:

```
root@beaglebone:~# route add default gw 192.168.7.1
```

Note that the eth1 device is the preferred Internet connection on my host system, while the eth4 device is the BeagleBone Black's device as viewed on my host system so you have to modify the command accordingly in order order to suite your needs.

Other conventions

New terms and **important words** are shown in bold. Words that you see on the screen, for example, in menus or dialog boxes, appear in the text like this: "Clicking the **Next** button moves you to the next screen."

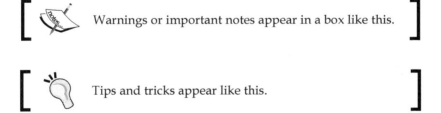

Warnings or important notes appear in a box like this.

Tips and tricks appear like this.

Reader feedback

Feedback from our readers is always welcome. Let us know what you think about this book—what you liked or disliked. Reader feedback is important for us as it helps us develop titles that you will really get the most out of.

To send us general feedback, simply e-mail feedback@packtpub.com, and mention the book's title in the subject of your message.

If there is a topic that you have expertise in and you are interested in either writing or contributing to a book, see our author guide at www.packtpub.com/authors.

Customer support

Now that you are the proud owner of a Packt book, we have a number of things to help you to get the most from your purchase.

Downloading the example code

You can download the example code files from your account at http://www.packtpub.com for all the Packt Publishing books you have purchased. If you purchased this book elsewhere, you can visit http://www.packtpub.com/support and register to have the files e-mailed directly to you.

For this book, the example code can also be downloaded from the author's GitHub repository at https://github.com/giometti/beaglebone_home_automation_blueprints.

Just use the following command to get it at once:

```
$ git clone https://github.com/giometti/beaglebone_home_automation_blueprints
```

The examples are grouped according to the chapter names, so you can easily find the code during the reading of the book.

Downloading the color images of this book

We also provide you with a PDF file that has color images of the screenshots/diagrams used in this book. The color images will help you better understand the changes in the output. You can download this file from http://www.packtpub.com/sites/default/files/downloads/BeagleBoneHomeAutomationBlueprints_ColoredImages.pdf.

Errata

Although we have taken every care to ensure the accuracy of our content, mistakes do happen. If you find a mistake in one of our books—maybe a mistake in the text or the code—we would be grateful if you could report this to us. By doing so, you can save other readers from frustration and help us improve subsequent versions of this book. If you find any errata, please report them by visiting `http://www.packtpub.com/submit-errata`, selecting your book, clicking on the **Errata Submission Form** link, and entering the details of your errata. Once your errata are verified, your submission will be accepted and the errata will be uploaded to our website or added to any list of existing errata under the Errata section of that title.

To view the previously submitted errata, go to `https://www.packtpub.com/books/content/support` and enter the name of the book in the search field. The required information will appear under the **Errata** section.

Piracy

Piracy of copyrighted material on the Internet is an ongoing problem across all media. At Packt, we take the protection of our copyright and licenses very seriously. If you come across any illegal copies of our works in any form on the Internet, please provide us with the location address or website name immediately so that we can pursue a remedy.

Please contact us at `copyright@packtpub.com` with a link to the suspected pirated material.

We appreciate your help in protecting our authors and our ability to bring you valuable content.

Questions

If you have a problem with any aspect of this book, you can contact us at `questions@packtpub.com`, and we will do our best to address the problem.

1
Dangerous Gas Sensors

In this chapter, we will learn how to use the **BeagleBone Black** to monitor some dangerous gases in a room, such as *CO, methane, LPG,* and so on, and then enabling an acoustic and visive alarm in case of danger. Also, by using a GSM module, the user will be able to send an SMS message to a predefined phone number to alert, for instance, a relative.

Also, the user will be able to control, log, and display the measured concentrations from the system console/terminal by using a command-line interface (this to keep the code simple).

We'll see how to build the circuitry to manage the sensors and how to get the gas concentration from them. Then, we'll take a look at how to manage a GSM module in order to send SMS messages.

The basic of functioning

In this project, our BeagleBone Black will periodically read the environmental data from the sensors, comparing them with user selectable ranges, and then generate an alarm in case one (or more) data read is out of that range.

The sensors will be connected to the BeagleBone Black ADCs with a dedicated circuitry, while the alarms will be enabled with dedicated GPIO lines. Then a GSM module will be connected to our BeagleBone Black's serial port in order to send other alarm messages via SMS.

Setting up the hardware

As just stated, all devices are connected with the BeagleBone Black, which is the real core of the system, as shown in the following screenshot:

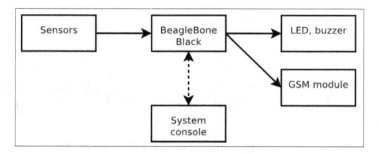

The data flow is from the sensors to the alarm actuators (LED, buzzer, and GSM module) and the user will be able to send commands to the system, or check the system status and the collected data, by using the system console.

Connecting the gas sensors

The gas sensors are used to monitor the environment and we can choose different kinds of such devices. I decided to use the ones shown in the following screenshot due to the fact they act as a variable resistor according to the gas concentration, so they can be easily read with a normal ADC:

In the prototype presented here, the gas sensors are actually four, but the ones named **MQ-2 (smoke detector)**, **MQ-4 (methane detector)**, and **MQ-7 (LPG detector)** look very similar to each other (except the label on each sensor), so I reported only one of them in the preceding screenshot, while the carbon monoxide detector is the red device labeled with MQ-7.

The devices can be purchased at the following links (or by surfing the Internet):

- MQ-2: http://www.cosino.io/product/mq-2-gas-sensor
- MQ-4: http://www.cosino.io/product/mq-4-gas-sensor
- MQ-5: http://www.cosino.io/product/mq-5-gas-sensor
- MQ-7: http://www.cosino.io/product/mq-7-gas-sensor

The following are the URLs where we can get the datasheet for each GAS sensor:

- MQ-2: http://www.seeedstudio.com/depot/datasheet/MQ-2.pdf
- MQ-4: https://www.pololu.com/file/0J311/MQ4.pdf
- MQ-5: http://www.dfrobot.com/image/data/SEN0130/MQ-5.pdf
- MQ-7: https://www.parallax.com/sites/default/files/downloads/605-00007-MQ-7-Datasheet.pdf

Looking carefully at the datasheet of the gas sensors, we can see exactly how these sensors' class varies their internal resistance according to the gas concentration (in reality, it depends on environment humidity and temperature too; but for an indoor functioning, we can consider these values as constants). So, if we put it in series with a resistor and apply a constant voltage). We can get an output voltage that is proportional to the actual gas concentration.

The following diagram shows a possible schematics where the gas sensor is connected to **5V** power supply and the **RL** resistor is formed by two resistors (**R1** & **R2**) due to the fact we cannot put more than 1.8V at a BeagleBone Black's ADC pin. So, by choosing these two resistors in such a way that $R1 \geq 2*R2$, we can be sure we have no more than $5.0V/3 \approx 1.67V$ at the ADC input pin on every possible functioning condition, even if the sensor's internal resistance is shorted. However, to be completely sure we can add a **Zener diode** (**Z**) with a reverse threshold on 1.8V (but I didn't use it on my prototype).

The following diagram shows the circuitry I used to connect each sensor:

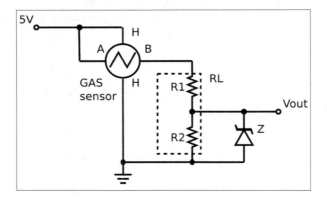

 Note that the **GAS sensors** have six pins labeled in pairs as **A**, **B**, and **H**; while the **A** and **B** pair pins are shortened, the **H** labeled pairs must be connected at one end to the input voltage (5V in our case) and the other end to the GND (see the datasheet for further information).

Another important issue regarding these sensors is the calibration we should perform before using them. This last adjustment is very important; as reported in the MQ-2 datasheet, we read the following recommendation:

> *We recommend that you calibrate the detector for 1000 ppm* **liquefied petroleum gas** *(LPG), or 1000ppm* **iso-butane** *(i-C$_4$H$_{10}$) concentration in air and use value of* **load resistance** *that (RL) about 20K (5K to 47K).*

This step can be done by replacing resistors **R1** or **R2** with a **varistor** and then fine tuning its resistance. However, I decided to use normal resistors (*R1 = 15KΩ, R2 = 6.8KΩ* in such a way that *RL = R1 + R2 ≈ 20KΩ*, as suggested by the datasheet) and then implemented a little translation in software (see the following section), that is, we can translate raw data from the ADCs into a **ppm** (**part-per-million**) value in such a way the user can work with physic data.

This translation can be done using a *gain* and an *offset* value in the following formula for each sensor:

- *ppm = raw * gain + offset*

During the calibration procedure, we just need to use two known points (*ppm1* and *ppm2*), read the corresponding raw data (*raw1* and *raw2*), and then apply the following formulas:

- *gain = (ppm1 – ppm2) / (raw1 – raw2)*
- *offset = ppm1 – raw1 * gain*

Of course, we need four gain/offset couples, one per sensor (the calibration procedure is quite long!)

Once we have fixed the input circuits, we simply have to connect each **Vout** to the BeagleBone Black's ADC input pins. Our board has 8 ADCs inputs, so we can use the following connections:

Pin	Gas sensor
P9.39 - AIN0	Vout @MQ-2
P9.37 - AIN2	Vout @MQ-4
P9.35 - AIN6	Vout @MQ-5
P9.33 - AIN4	Vout @MQ-7

To enable them, we use the following command:

```
root@beaglebone:~# echo cape-bone-iio > /sys/devices/bone_capemgr.9/slots
```

If everything works well, we should get the following kernel messages:

```
part_number 'cape-bone-iio', version 'N/A'
slot #7: generic override
bone: Using override eeprom data at slot 7
slot #7: 'Override Board Name,00A0,Override Manuf,cape-bone-iio'
slot #7: Requesting part number/version based 'cape-bone-iio-00A0.dtbo'
slot #7: Requesting firmware 'cape-bone-iio-00A0.dtbo' for board-name
'Override Board Name', version '00A0'
slot #7: dtbo 'cape-bone-iio-00A0.dtbo' loaded; converting to live tree
slot #7: #1 overlays
helper.12: ready
slot #7: Applied #1 overlays.
```

Then, the files AIN0, AIN1, ..., AIN7 should become available as follows:

```
root@beaglebone:~# find /sys -name '*AIN*'
/sys/devices/ocp.3/helper.12/AIN0
/sys/devices/ocp.3/helper.12/AIN1
/sys/devices/ocp.3/helper.12/AIN2
/sys/devices/ocp.3/helper.12/AIN3
/sys/devices/ocp.3/helper.12/AIN4
/sys/devices/ocp.3/helper.12/AIN5
/sys/devices/ocp.3/helper.12/AIN6
/sys/devices/ocp.3/helper.12/AIN7
```

 These settings can be done using the bin/load_firmware.sh script in the book's example code repository, as follows:

```
root@beaglebone:~# ./load_firmware.sh adc
```

Then, we can read the input data by using the cat command:

```
root@beaglebone:~# cat /sys/devices/ocp.3/helper.12/AIN0
1716
```

 Note that the ADC can also be read by using other files still into the **sysfs** filesystem. The following command, for instance, reads from AIN0 input pin:

```
root@beaglebone:~# cat /sys/bus/iio/devices/
iio:device0/in_voltage0_raw
```

Connecting the alarm actuators

Now, we have to connect the alarm actuators in such a way the user can have a visual and acoustic feedback of any possible dangerous gas concentration. Also, we have to connect the GSM module to a serial port to communicate with it.

LED and buzzer

The LED and buzzer connections are very simple. The LEDs can be directly connected (with a resistor) with the BeagleBone Black's GPIO pins without problems, while the buzzer needs a little more work due to the fact that it needs a higher current than the LED to work. However, we can resolve the problem by using a transistor as shown in the following diagram to manage the buzzer with a higher current.

Note that the buzzer can't be a simple piezo without an internal oscillator, otherwise an external oscillator circuit or a **PWM** signal must be used!

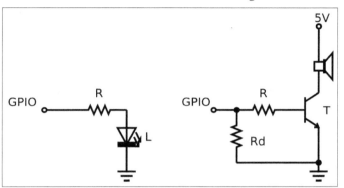

In my circuitry, I used an **R** (470Ω) resistor for the **LED (L)**, an **R** (2KΩ), **Rd** (4.7KΩ) resistors for the buzzer, and a **BC546 transistor (T)**. Note that, regarding the LEDs, an R = 100Ω resistor can result in a higher brightness, so you may change it according to the LED color to have different results.

Note also that the resistor **Rd** in the buzzer circuitry is needed to pull-down the GPIO during the boot. In fact, during this stage it is set as input, and even in such configuration the current that flows out from the pin can turn on the buzzer!

The BeagleBone Black has a lot of GPIOs lines, so we can use the following connections:

Pin	Actuator
P8.9 - GPIO69	R @LED
P8.10 - GPIO68	R @Buzzer

Now, to test the connections, we can set up the **GPIO**s by exporting them and then setting these lines as outputs with the following commands:

```
root@beaglebone:~# echo 68 > /sys/class/gpio/export
root@beaglebone:~# echo out > /sys/class/gpio/gpio68/direction
root@beaglebone:~# echo 0 > /sys/class/gpio/gpio68/value
root@beaglebone:~# echo 69 > /sys/class/gpio/export
root@beaglebone:~# echo out > /sys/class/gpio/gpio69/direction
root@beaglebone:~# echo 0 > /sys/class/gpio/gpio69/value
```

 Note that it will be a good idea to use blinking LEDs to do this job. However, for this first chapter I'm going to use normal GPIO lines, leaving this topic for the following chapters.

Now, to turn on and off both the LED and the buzzer, we simply write 1 or 0 into the proper files, as follows:

```
root@beaglebone:~# echo 1 > /sys/class/gpio/gpio68/value
root@beaglebone:~# echo 0 > /sys/class/gpio/gpio68/value
root@beaglebone:~# echo 1 > /sys/class/gpio/gpio69/value
root@beaglebone:~# echo 0 > /sys/class/gpio/gpio69/value
```

 These settings can be done by using the bin/gpio_set.sh script in the book's example code repository, as follows:

```
root@beaglebone:~# ./gpio_set 68 out
root@beaglebone:~# ./gpio_set 69 out
```

GSM module

As stated in the introduction of this chapter, we wish to add a GSM module to be able to alert the user remotely too. In order to do this, we can connect this device with a normal serial port with TTL level signals. In this case, we have only to choose one of the serial ports available on our BeagleBone Black.

The following screenshot shows the GSM module I decided to use:

  The device can be purchased at the following link (or by surfing the Internet):

http://www.cosino.io/product/serial-gsmgprs-module

The user manual con be retrieved at http://www.mikroe.com/downloads/get/1921/gsm_click_manual_v101c.pdf.

The BeagleBone Black has four available serial ports. By deciding to use the device /dev/ttyO1, we can use the following connections:

Pin	GSM module
P9.24 - TX-O1	RX
P9.26 - RX-O1	TX
P9.1 - GND	GND
P9.3 - 3.3V	3.3V
P9.5 - 3.3V	5V

To enable the **serial port**, we have to use the following command:

```
root@beaglebone:~# echo BB-UART1 > /sys/devices/bone_capemgr.9/slots
```

If everything works well, we should get the following kernel messages:

```
part_number 'BB-UART1', version 'N/A'
slot #8: generic override
bone: Using override eeprom data at slot 8
slot #8: 'Override Board Name,00A0,Override Manuf,BB-UART1'
slot #8: Requesting part number/version based 'BB-UART1-00A0.dtbo
slot #8: Requesting firmware 'BB-UART1-00A0.dtbo' for board-name
'Override Board Name', version '00A0'
slot #8: dtbo 'BB-UART1-00A0.dtbo' loaded; converting to live tree
slot #8: #2 overlays
48022000.serial: ttyO1 at MMIO 0x48022000 (irq = 73) is a OMAP UART1
slot #8: Applied #2 overlays.
```

The device file /dev/ttyO1 should now become available.

These settings can be done by using the bin/load_firmware.sh script in the book's example code repository, as follows:

```
root@beaglebone:~# ./load_firmware.sh ttyO1
```

To verify that the new device is ready, we can use the ls command as follows:

```
root@beaglebone:~# ls -l /dev/ttyO1
crw-rw---T 1 root dialout 248, 1 Apr 23 22:25 /dev/ttyO1
```

The reader can take a look at the book *BeagleBone Essentials, Packt Publishing,* which was written by the author of this book, in order to have more information regarding how to activate and use the GPIO lines and the serial ports available on the system.

Now, we can test whether we actually talk with the modem by using the screen command as follows:

```
root@beaglebone:~# screen /dev/ttyO1 115200
```

The screen command can be installed by using the aptitude command as follows:

```
root@beaglebone:~# aptitude install screen
```

After pressing the *ENTER* key, you should get a blank terminal where, if you enter the ATZ string, you should get the string OK as answer, as shown in the following code:

ATZ

OK

It's the GSM module that answers that it's okay and fully functional. To quit from the screen command, you have to enter the *CTRL* + *A* + \ keys sequence and then answer *yes* by pressing the *y* key when the program asks you to Really quit and kill all your windows [y/n].

The final picture

Well, now we have to put it all together! The following image shows the prototype I made to implement this project and to test the software:

Note that we need an external power supplier due to the fact that the external circuitry (and especially the GSM module) needs the 5V power supply.

Setting up the software

Now, it's time to think about the software needed to implement the desired functioning, that is, checking the gas concentrations, logging them, and eventually activating the alarms. We need the following:

1. A periodic procedure (`read_sensors.php`) that periodically scans all the sensors and then logs their data into a database.

2. A periodic procedure (`monitor.php`) that reads the sensors' data, checks them against preset thresholds, and then sets some internal status.

3. A periodic procedure (`write_actuators.php`) that enables the alarms according to the previously saved status.

The following diagram shows the situation:

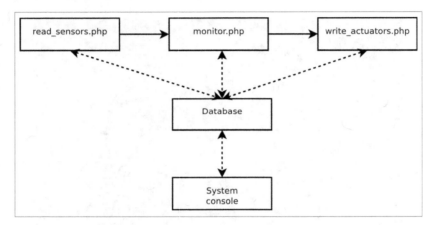

The core of the system is the database, where we store both the data we wish to log and the system's status. In this manner, all periodic functions can be realized as separate tasks that talk to each other by using the database itself. Also, we can control all the tasks from the system console by just altering the `config` table at runtime.

I used **MySQL** to implement the database system, and the preceding configuration can be created by using the `my_init.sh` script, where we define the proper tables.

The MySQL daemon can be installed by using the `aptitude` command as follows:

root@beaglebone:~# aptitude install mysql-client mysql-server

Here is a snippet of the script:

```
CREATE TABLE status (
    n VARCHAR(64) NOT NULL,
    v VARCHAR(64) NOT NULL,
    PRIMARY KEY (n)
) ENGINE=MEMORY;

# Setup default values
INSERT INTO status (n, v) VALUES('alarm', 'off');

#
# Create the system configuration table
#

CREATE TABLE config (
    n VARCHAR(64) NOT NULL,
    v VARCHAR(64) NOT NULL,
    PRIMARY KEY (n)
);

# Setup default values
INSERT INTO config (n, v) VALUES('sms_delay_s', '300');

INSERT INTO config (n, v) VALUES('mq2_gain', '1');
INSERT INTO config (n, v) VALUES('mq4_gain', '1');
INSERT INTO config (n, v) VALUES('mq5_gain', '1');
INSERT INTO config (n, v) VALUES('mq7_gain', '1');
INSERT INTO config (n, v) VALUES('mq2_off', '0');
INSERT INTO config (n, v) VALUES('mq4_off', '0');
INSERT INTO config (n, v) VALUES('mq5_off', '0');
INSERT INTO config (n, v) VALUES('mq7_off', '0');

INSERT INTO config (n, v) VALUES('mq2_th_ppm', '2000');
INSERT INTO config (n, v) VALUES('mq4_th_ppm', '2000');
INSERT INTO config (n, v) VALUES('mq5_th_ppm', '2000');
INSERT INTO config (n, v) VALUES('mq7_th_ppm', '2000');
```

```
#
# Create one table per sensor data
#

CREATE TABLE MQ2_log (
    t DATETIME NOT NULL,
    d float,
    PRIMARY KEY (t)
);

CREATE TABLE MQ4_log (
    t DATETIME NOT NULL,
    d float,
    PRIMARY KEY (t)
);

CREATE TABLE MQ5_log (
    t DATETIME NOT NULL,
    d float,
    PRIMARY KEY (t)
);

CREATE TABLE MQ7_log (
    t DATETIME NOT NULL,
    d float,
    PRIMARY KEY (t)
);
```

 The `my_init.sh` script is stored in the `chapter_01/my_init.sh` file in the book's example code repository.

The reader should notice that we define a `status` table with the MEMORY storage engine since we don't need to preserve it at reboot but need a good performance in accessing it, while the `config` table and the per-sensor logging tables (`MQ2_log`, `MQ4_log`, `MQ5_log`, and `MQ7_log`) are defined as normal tables since we need to save these data even during a complete restart. Note that we defined one table per variable in order to easily get access to the logged data; however, nothing changes, even if we decide to keep the logged data into a global logging table.

Note also that during the database initialization, we can define some default settings by simply recording these values by using an INSERT command. For the status table, we just need the alarm variable to be set to off, while into the config table, we can set up the minimum delay in seconds (sms_delay_s) to wait before resending a new SMS alarm, the gain/offset translation couple variables (mq2_gain/ mq2_off and friends), and the per-sensor threshold variables (mq2_th_ppm and friends) to be used to activate the alarms.

Managing the ADCs

Now, to get data from the **ADC** and save them into the database, we have to write a periodic task. This is quite easy and the following code snippet shows a PHP implementation of the main function of the file read_sensors.php, which does this:

```php
function daemon_body()
{
    global $loop_time;
    global $sensors;

    # The main loop
    dbg("start main loop (loop_time=${loop_time}s)");
    while (sleep($loop_time) == 0) {
        dbg("loop start");

        # Read sensors
        foreach ($sensors as $s) {
            $name = $s['name'];
            $file = $s['file'];
            $var = $s['var'];
            $log = $s['log'];

            # Get the converting values
            $gain = db_get_config($var . "_gain");
            $off = db_get_config($var . "_off");

            dbg("gain[$var]=$gain off[$var]=$off");

            # Read the ADC file
            $val = file_get_data($file);
            if ($val === false) {
                err("unable to read sensor $name");
                continue;
            }
```

```
        # Do the translation
        $ppm = $val * $gain + $off;

        dbg("file=$file val=$val ppm=$ppm");

        # Store the result into the status table
        $ret = db_set_status($var, $ppm);
        if (!$ret) {
            err("unable to save $name status db_err=%s",
                mysql_error());
            continue;
        }

        # Store the result into the proper log table
        $ret = db_log_var($log, $ppm);
        if (!$ret)
            err("unable to save $name log db_err=%s",
                mysql_error());
        }

        dbg("loop end");
    }
}
```

 The complete script is stored in the chapter_01/read_sensors.php file in the book's example code repository.

The function is quite simple. It starts the main loop to periodically read the ADC data, get the *gain* and *offset* conversion values for the current variable needed to convert it into the corresponding *ppm* number, then alters the current status variables, and adds a new value into the logging table of the read sensor.

If we execute the script enabling all debugging command-line options, we get:

```
root@beaglebone:~# ./read_sensors.php -d -f -l -T 5
read_sensors.php[5388]: signals traps installed
read_sensors.php[5388]: start main loop (loop_time=5s)
read_sensors.php[5388]: loop start
read_sensors.php[5388]: gain[mq2]=0.125 off[mq2]=0
read_sensors.php[5388]: file=/sys/devices/ocp.3/helper.12/AIN0 val=810
ppm=101.25
read_sensors.php[5388]: gain[mq4]=1 off[mq4]=0
```

```
read_sensors.php[5388]: file=/sys/devices/ocp.3/helper.12/AIN2 val=1477
ppm=1477

read_sensors.php[5388]: gain[mq5]=1 off[mq5]=0

read_sensors.php[5388]: file=/sys/devices/ocp.3/helper.12/AIN6 val=816
ppm=816

read_sensors.php[5388]: gain[mq7]=1 off[mq7]=0

read_sensors.php[5388]: file=/sys/devices/ocp.3/helper.12/AIN4 val=572
ppm=572

read_sensors.php[5388]: loop end

read_sensors.php[5388]: loop start

read_sensors.php[5388]: gain[mq2]=0.125 off[mq2]=0

read_sensors.php[5388]: file=/sys/devices/ocp.3/helper.12/AIN0 val=677
ppm=84.625

read_sensors.php[5388]: gain[mq4]=1 off[mq4]=0

read_sensors.php[5388]: file=/sys/devices/ocp.3/helper.12/AIN2 val=1456
ppm=1456

read_sensors.php[5388]: gain[mq5]=1 off[mq5]=0

read_sensors.php[5388]: file=/sys/devices/ocp.3/helper.12/AIN6 val=847
ppm=847

read_sensors.php[5388]: gain[mq7]=1 off[mq7]=0

read_sensors.php[5388]: file=/sys/devices/ocp.3/helper.12/AIN4 val=569
ppm=569

read_sensors.php[5388]: loop end

...
```

 Note that only the first sensor has been (more or less) calibrated!

The process can be stopped as usual with the *CTRL + C* sequence.

Now, we can read the system status (in this case, the last read sensors datum) by using the my_dump.sh script, as follows:

```
root@beaglebone:~# ./my_dump.sh status
n      v
alarm    off
mq2    84.625
mq4    1456
mq5    815
mq7    569
```

9

 The `my_dump.sh` script is stored in the `chapter_01/my_dump.sh` file in the book's example code repository.

The same script can be used to dump a logging table too. For instance, if we wish to see the MQ-2 logged data, we can use the following command:

```
root@beaglebone:~# ./my_dump.sh mq2_log
t      v
2015-05-15 17:39:36 101.25
2015-05-15 17:39:41 84.625
2015-05-15 17:39:46 84.625
```

Managing the actuators

When a sensor detects a dangerous gas concentration, the `alarm` status variable is set to *on* state. Therefore, when this happens, we have to turn both the LED and the buzzer on, and we must send an SMS message to the user's predefined number.

In order to do these actions, we have to properly set up the GPIO lines that manage the LED and the buzzer as shown previously, and then we have to talk with the **GSM** module through the serial port to send the SMS message. To do this last step, we have to to install the `gsm-utils` package where we can find the `gsmsendsms` command, which is used to actually send the SMS. In order to install the package, we use the following command:

```
root@beaglebone:~# aptitude install gsm-utils
```

Then, after placing a functioning SIM into the module, we can verify to be able to talk with the GSM module with the `gsmctl` command, as shown in the following code:

```
root@beaglebone:~# gsmctl -d /dev/ttyO1 me
<ME0>  Manufacturer: Telit
<ME1>  Model: GL865-QUAD
<ME2>  Revision: 10.00.144
<ME3>  Serial Number: 356308042878501
```

Then, we can verify the current PIN status by using the following command:

```
root@beaglebone:~# gsmctl -d /dev/ttyO1 pin
<PIN0> READY
```

The preceding message shows us that the GSM module is correctly configured and the SIM in it is ready to operate; however, the SIM must be enabled by inserting the proper PIN number if we get the following message:

```
gsmsendsms[ERROR]: ME/TA error 'SIM PIN required' (code 311)
```

In this case, we must use the following command:

```
root@beaglebone:~# gsmctl -d /dev/ttyO1 -I "+cpin=NNNN"
```

In the preceding command, NNNN is the PIN number of your SIM. If the command hangs with no output at all, it means that the connection is wrong.

Now that we've checked the connection and the SIM is enabled, we can start to send SMS messages by using the following command:

```
root@beaglebone:~# gsmsendsms -d /dev/ttyO1 "+NNNNNNNNNNNNN" 'Hello
world!'
```

In the preceding command, the NNNNNNNNNNNNN string is the number where the SMS must be sent.

> If the module answers is as follows it means that **SMS Service Centre Address (SCA)**; which is the phone number of the centre that is accepting SMS for delivery is not set correctly in your phone:
>
> ```
> gsmsendsms[ERROR]: ME/TA error 'Unidentified
> subscriber' (code 28)
> ```
>
> In this case, you should ask to your GSM operator and then try the following command:
>
> ```
> root@beaglebone:~# gsmctl -o setsca "+SSSSSSSSSSSS"
> ```
>
> In the preceding command, the SSSSSSSSSSSS string is the number of your centre.

Okay, now we have all the needed information to control our actuators. A possible implementation of main function of the managing task is as follows:

```
function daemon_body()
{
    global $loop_time;
    global $actuators;

    $sms_delay = db_get_config("sms_delay_s");
```

```php
$old_alarm = 0;
$sms_time = strtotime("1970");

# The main loop
dbg("start main loop (loop_time=${loop_time}s)");
while (sleep($loop_time) == 0) {
dbg("loop start");

    # Get the "alarm" status and set all alarms properly
    $alarm = db_get_status("alarm");
        foreach ($actuators as $a) {
            $name = $a['name'];
            $file = $a['file'];

            dbg("file=$file alarm=$alarm");
            $ret = gpio_set($file, $alarm);
                if (!$ret)
                    err("unable to write actuator $name");
        }

    # Send the SMS only during off->on transition
    if ($alarm == "on" && $old_alarm == "off" &&
        strtotime("-$sms_time seconds") > $sms_delay) {
            do_send_sms();
            $sms_time = strtotime("now");
        }

    $old_alarm = $alarm;

    dbg("loop end");
    }
}
```

 The complete script is stored in the `chapter_01/write_actuators.php` file in the book's example code repository.

Again, the function is really simple — we simply have to read the current `alarm` variable status from the database and then set up the actuators according to it. Note that a special job must be done for the SMS management; in fact, the system must send one SMS at time and only during the *off-to-on* transition and not before `sms_delay` seconds. To do the trick, we use the `old_alarm` and `sms_time` variables to save the last loop status.

To test the code, we can control the `alarm` variable by using the `my_set.sh` command as follows:

```
root@beaglebone:~# ./my_set.sh status alarm on
root@beaglebone:~# ./my_set.sh status alarm off
```

 The script is stored in the `chapter_01/my_set.sh` file in the book's example code repository.

So, let's start the script with the command:

```
root@beaglebone:~# ./write_actuators.php -d -f -l -T 5
write_actuators.php[5474]: signals traps installed
write_actuators.php[5474]: start main loop (loop_time=5s)
write_actuators.php[5474]: loop start
write_actuators.php[5474]: file=/sys/class/gpio/gpio68 alarm=off
write_actuators.php[5474]: file=/sys/class/gpio/gpio69 alarm=off
write_actuators.php[5474]: loop end
write_actuators.php[5474]: loop start
write_actuators.php[5474]: file=/sys/class/gpio/gpio68 alarm=off
write_actuators.php[5474]: file=/sys/class/gpio/gpio69 alarm=off
write_actuators.php[5474]: loop end
```

On another terminal, we can change the `alarm` variable, as already stated, by using the following command:

```
root@beaglebone:~# ./my_set.sh status alarm on
```

After this we notice that the script does its job:

```
write_actuators.php[5474]: loop start
write_actuators.php[5474]: file=/sys/class/gpio/gpio68 alarm=on
write_actuators.php[5474]: file=/sys/class/gpio/gpio69 alarm=on
write_actuators.php[5474]: send SMS...
write_actuators.php[5474]: loop end
```

Regarding how to send an SMS message in PHP, I simply used the following code:

```
function do_send_sms()
{
   dbg("send SMS...");
   system('gsmsendsms -d /dev/ttyO1 "' . PHONE_NUM . '" "GAS alarm!"');
}
```

Basically, here we use the `system()` function to call the `gsmsendsms` command.

 You may note that `gsmsendsms` takes a while to send the SMS. It's normal.

Controlling the environment

Now, we only need the glue between the sensors and actuators managing tasks, that is, a periodic function that according to the user inputs periodically checks whether the alarms must be activated according to the information read, or not.

A possible implementation of the main function of the `monitor.php` script is as follows:

```
function daemon_body()
{
   global $loop_time;
   global $actuators;

   # The main loop
   dbg("start main loop (loop_time=${loop_time}s)");
   while (sleep($loop_time) == 0) {
      dbg("loop start");

      # Get the gas concentrations and set the "alarm" variable
      $mq2 = db_get_status("mq2");
      $mq2_th_ppm = db_get_config("mq2_th_ppm");
      dbg("mq2/mq2_th_ppm=$mq2/$mq2_th_ppm");
      $mq4 = db_get_status("mq4");
      $mq4_th_ppm = db_get_config("mq4_th_ppm");
      dbg("mq4/mq4_th_ppm=$mq4/$mq4_th_ppm");
      $mq5 = db_get_status("mq5");
      $mq5_th_ppm = db_get_config("mq5_th_ppm");
      dbg("mq5/mq5_th_ppm=$mq5/$mq5_th_ppm");
      $mq7 = db_get_status("mq7");
```

```
    $mq7_th_ppm = db_get_config("mq7_th_ppm");
    dbg("mq7/mq7_th_ppm=$mq7/$mq7_th_ppm");

    $alarm = $mq2 >= $mq2_th_ppm ||
        $mq2 >= $mq2_th_ppm ||
        $mq2 >= $mq2_th_ppm ||
        $mq2 >= $mq2_th_ppm ? 1 : 0;

    db_set_status("alarm", $alarm);
    dbg("alarm=$alarm");

    dbg("loop end");
    }
}
```

 The complete script is stored in the chapter_01/monitor.php file in the book's example code repository.

The function starts the main loop where, after getting the sensors' thresholds, it simply gets the last sensor's values and sets up the alarm variable accordingly.

Again, we can change the gas concentration thresholds by using the my_set.sh command as follows:

root@beaglebone:~# ./my_set.sh config mq2_th_ppm 5000

We can test the script by executing it in the same manner as the previous two, as follows:

root@beaglebone:~# ./monitor.php -d -f -l -T 5
monitor.php[5819]: signals traps installed
monitor.php[5819]: start main loop (loop_time=5s)
monitor.php[5819]: loop start
monitor.php[5819]: mq2/mq2_th_ppm=84.625/5000
monitor.php[5819]: mq4/mq4_th_ppm=1456/2000
monitor.php[5819]: mq5/mq5_th_ppm=815/2000
monitor.php[5819]: mq7/mq7_th_ppm=569/2000
monitor.php[5819]: alarm=0
monitor.php[5819]: loop end
monitor.php[5819]: loop start
monitor.php[5819]: mq2/mq2_th_ppm=84.625/5000
monitor.php[5819]: mq4/mq4_th_ppm=1456/2000

```
monitor.php[5819]: mq5/mq5_th_ppm=815/2000
monitor.php[5819]: mq7/mq7_th_ppm=569/2000
monitor.php[5819]: alarm=0
monitor.php[5819]: loop end
...
```

To stop the test, just use the *CTRL + C* sequence. You should get an output as follows:

```
^Cmonitor.php[5819]: signal trapped!
```

Final test

Once everything has been connected and the software is ready, it's time to do a little test of our new system. The demonstration can be done by using a lighter. In fact, our system is really sensitive to the gas inside the lighter!

First of all, we have to check the system configuration:

```
root@beaglebone:~# ./my_dump.sh config
n      v
mq2_gain    0.125
mq2_off     0
mq2_th_ppm      150
mq4_gain    0.125
mq4_off     0
mq4_th_ppm      150
mq5_gain    0.125
mq5_off     0
mq5_th_ppm      150
mq7_gain    0.125
mq7_off     0
mq7_th_ppm      150
sms_delay_s     300
```

 Note that I used a very weak calibration setting; however, these are okay for a demo.

Then, we can take a look at the system's current status:

```
root@beaglebone:~# ./my_dump.sh status
n     v
mq2   73.5
mq4   121.75
mq5   53
mq7   80.5
alarm  0
```

Then, we can do all hardware settings at once by using the chapter_01/SYSINIT.sh script in the book's example code repository as follows:

```
root@beaglebone:~# ./SYSINIT.sh
done!
```

Okay, now let's start all the required process daemons:

```
root@beaglebone:~# ./read_sensors.php -d -T 2
root@beaglebone:~# ./write_actuators.php -d -T 2
root@beaglebone:~# ./monitor.php -d -T 2
```

> Note that all the daemons are running in background in this way; however, the debugging messages are enabled and they can be viewed into the system log with the following command:
> ```
> # tail -f /var/log/syslog
> ```

Now, we have to approach the lighter to the sensors and press the button on the lighter in order to allow the sensor to detect the gas. After a while, the alarms should be turned on, and looking at the system status, we should get the following:

```
root@beaglebone:~# ./my_dump.sh status
n     v
mq2   203.875
mq4   166.5
mq5   52.5
mq7   122.625
alarm  1
```

Also, if we have set up a phone number, we should receive an SMS on the phone!

As last step, let's display the data logged by plotting them. We can use the following command to extract the data from the database:

```
root@beaglebone:~# ./my_dump.sh mq2_log | awk '{ print $2 " " $3 }' >
mq2.log
```

In the mq2.log file, we should find something like the following:

```
root@beaglebone:~# cat mq2.log
15:02:07 75.25
15:02:10 74.25
15:02:12 74.25
15:02:14 74.375
15:02:16 74.25
...
```

Now, using the next command, we're going to create a PNG image holding a plot of our data:

```
$ gnuplot mq2.plot
```

Note that in order to execute this command, you need the gnuplot command, which can be installed by using the following command:

```
# aptitude install gnuplot
```

Also, both the mq2.log and mq2.plot files are need. The former is created by the preceding command line, while the latter can be found in the chapter_01/mq2.plot file in the book's example code repository. It holds the gnuplot instructions to effectively draw the plot.

The plot of the MQ-2 data of my test is shown in the following screenshot:

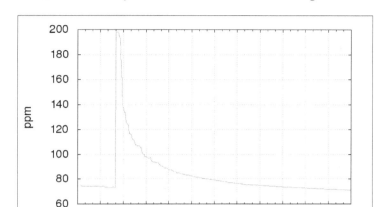

As you can see, the sensors are very sensitive to the gas; as soon as I opened my lighter and the gas reached them, the ppm concentration went to high values very quickly.

To stop the test, we can use the following commands:

```
root@beaglebone:~# killall read_sensors.php
root@beaglebone:~# killall write_actuators.php
root@beaglebone:~# killall monitor.php
```

Summary

In this chapter, we discovered how to manage the ADCs, the GPIOs lines, a GSM module thought, and a serial port. We also saw how to implement a simple monitoring program that can communicate (through a database server) with a sensors reader task to collect input data, and with an actuators manager to alert the user in case of emergency.

In the next chapter, we'll see how to manage an **ultrasonic distance** sensor to implement a parking assistant that can communicate to the driver, the distance between the car and the garage's wall. However, the really interesting part of the next chapter is about how to manage the distance sensor in two different setups: one with all the peripherals near the BeagleBone Black, and another with a remote connection of the sensor through a USB cable.

Ultrasonic Parking Assistant

2

In this chapter, we'll learn how to use the BeagleBone Black to implement a park assistant. We're going to use an ultrasonic sensor to detect the distance between our car and the garage wall, and some LEDs to give a feedback of the car position to the driver in order to avoid collisions.

We'll see how to set up the ultrasonic range sensor in two different manners, using different interfaces to get the data in order to resolve a problem in two different ways and obtain two different system configurations.

The basic of functioning

The project is really simple, even if it needs some electronic skills in order to manage the sensor output. Basically, our BeagleBone Black just needs to periodically poll the ultrasonic sensor output and then turn on the LEDs according to the distance from the wall: as level indicator lower is the distance and more LEDs are turned on.

Setting up the hardware

As just stated, in this project we're trying to implement two different setups: the first one uses the analog output of the ultrasonic sensor and implements a circuitry, where all the devices are directly connected with the BeagleBone Black (all peripherals are near the board); on the other hand, the second setup allows us to remotely manage the ultrasonic sensor by using an USB connection, so we can mount the sensor far from the BeagleBone Black board.

Simply speaking, we can put the sensor in one place while the LEDs are in a different location, maybe in a more visible position, as shown in the following image:

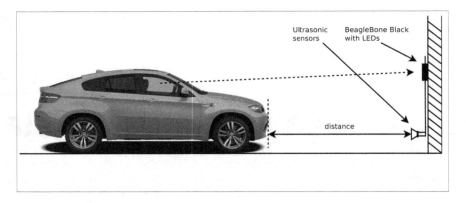

As you can see, the dotted arrow, which represents the driver's point of view, is more clear if the LEDs are in a upper position with respect to the distance sensor that should be located near to the floor to better catch the car frontal.

First setup – all devices near the BeagleBone Black

In this setup, we're going to use an ADC pin of our BeagleBone Black to read the analog output of the ultrasonic sensor.

Using the analog output of the distance sensor

The following image shows the ultrasonic sensor that I used on my prototype:

The devices can be purchased at the following link (or by surfing the Internet):

`http://www.cosino.io/product/ultrasonic-distance-sensor`.

The datasheet of this device is available at `http://www.maxbotix.com/documents/XL-MaxSonar-EZ_Datasheet.pdf`.

This device is really interesting due to the fact it has several output channels useful to retrieve the measured distance. In particular, it can give us the measurement via an analog voltage channel and via a serial port; the former communication channel is used in this setup, while the latter will be discussed in the second setup.

Looking into the datasheet, we discover that the analog output has a resolution of *Vcc/1024 per cm* with a maximum reported range of ~700 mm at 5V and ~600 cm at 3.3V. In this setup, we use Vcc set to 3.3V so the maximum output voltage (**VoutMAX**) will result as:

- $VoutMAX = 3.3V / 1024 * 600 \approx 1.93V$

Remembering that the BeagleBone Black's ADCs have a maximum input voltage of 1.8V, we have to find a way to scale down this value. A *quick and dirty* trick is to use a classic voltage divider, as shown in the following diagram:

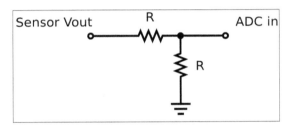

By using the preceding circuit, we simply divide the sensor output by 2. The voltage at ADC in pin is given by the following formula:

- $V_{ADCin} = R / (R + R) * Vout = R / 2R * Vout = 1 / 2 * Vout$

So, the only thing to do is to choose a suitable value for the two resistors (**R**). In my prototype, I set this value to $R=6.8K\Omega$, which is a reasonable value to have to acquire a suitable current flooding out from the sensor.

In this situation, our resolution becomes ~1.61mV/cm, and the connections to be done on the BeagleBone Black are shown in the following table:

Pin	Distance sensor pin (label)
P9.1 - GND	7
P9.3 - 3.3V	6 (Vcc)
P9.39 - AIN0	3 (AN)

Now, to enable the BeagleBone Black's ADC lines, we can use the following command as we already did in *Chapter 1, Dangerous Gas Sensors*:

```
root@beaglebone:~# echo cape-bone-iio > /sys/devices/bone_capemgr.9/slots
```

If everything works well, we should get the following kernel messages:

```
part_number 'cape-bone-iio', version 'N/A'

slot #7: generic override

bone: Using override eeprom data at slot 7

slot #7: 'Override Board Name,00A0,Override Manuf,cape-bone-iio'

slot #7: Requesting part number/version based 'cape-bone-iio-00A0.dtbo'

slot #7: Requesting firmware 'cape-bone-iio-00A0.dtbo' for board-name 'Override Board Name', version '00A0'

slot #7: dtbo 'cape-bone-iio-00A0.dtbo' loaded; converting to live tree

slot #7: #1 overlays

bone-iio-helper helper.12: ready

slot #7: Applied #1 overlays.
```

Then, the `AIN0`, `AIN1`, ..., `AIN7` files should become available, as follows:

```
root@beaglebone:~# find /sys -name '*AIN*'

/sys/devices/ocp.3/helper.12/AIN0

/sys/devices/ocp.3/helper.12/AIN1

/sys/devices/ocp.3/helper.12/AIN2

/sys/devices/ocp.3/helper.12/AIN3

/sys/devices/ocp.3/helper.12/AIN4

/sys/devices/ocp.3/helper.12/AIN5

/sys/devices/ocp.3/helper.12/AIN6

/sys/devices/ocp.3/helper.12/AIN7
```

These settings can be done by using the `bin/load_firmware.sh` script in the book's example code repository, as follows:

```
root@beaglebone:~# ./load_firmware.sh adc
```

Then, we can read the input data by using the `cat` command, as follows:

```
root@beaglebone:~# cat /sys/devices/ocp.3/helper.12/AIN0
1716
```

 As already stated in *Chapter 1, Dangerous Gas Sensors*, the ADC can also be read by using another file's still into the *sysfs* filesystem with the following command:

```
root@beaglebone:~# cat /sys/bus/iio/devices/iio:device0/
in_voltage0_raw
```

Now, we have to find a way to convert the read values from the ADC into a distance measured in meters so that we can decide how to manage the LEDs to give the feedback to the driver. Recalling what was just said, the resolution is ~1.61mV/cm, and considering that the resolution of the ADC is 12 bits and the maximum voltage is 3.3V, the **distance (d)** in centimeters between the car and the wall is given by the following formula (where the value *n* is the data read from the ADC):

- $d = 3.3V * n / 4095 / 0.00161V/cm$

Note that these are estimated values, so it is better to do a calibration of the sensor in order to have the correct reads at least near the lowest value that we wish to measure (in our example, this value is 0.20 m.) To do this, we can put something at 20 cm from the sensor, measure the output value from the ADC, and then calculate a compensating value *K* in order that the following formula will return exactly the value 20:

- $d_{calib} = K * 3.3V * n/4095 / 0.00161V/cm$

Note that in case of no calibration, *K* can be set to `1` (In this case, we obtain again the original formula, $d = d_{calib}$.)

On my prototype, putting an object at 20 cm from the sensor, I get the following value:

```
root@beaglebone:~# cat /sys/devices/ocp.3/helper.12/AIN0
29
```

So, *K* should be set to `1.38`.

Connecting the LEDs in the first setup

The LEDs' connections are very simple since they can be directly connected with the BeagleBone Black's GPIO pins, as shown the following diagram, which shows the schematic of one single LED connection that can be replicated for each LED:

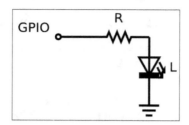

 I used an $R = 470\Omega$ resistor for the **LED** (**L**). Again, as in the previous chapter, let's remember that the resistor value **R** should be changed according to LED color if we wish to have a lighter effect.

We have 5 LEDs, so we need 5 GPIO lines. We can use the following connections:

Pin	LED color	Activated when distance is under
P8.45-GPIO44	White	5.00 m
P8.46-GPIO67	Yellow	2.00 m
P8.7-GPIO69	Red	1.00 m
P8.8-GPIO68	Red	0.50 m
P8.9-GPIO45	Red	0.20 m

The white LED is used to signal to the user that they are driving at less than 5 m from the wall; the yellow is used to signal that they are driving at less than 2 m from the wall; and the red LED is used to signal that the garage wall is approaching at less than 1 m, 0.50 m, and 0.20 m.

To test the LEDs' connection, we can use the same commands used in *Chapter 1, Dangerous Gas Sensors*. For instance, we can test the LED on GPIO68 by using the following commands to set up the GPIO first and then to turn it off and on:

```
root@beaglebone:~# echo 68 > /sys/class/gpio/export
root@beaglebone:~# echo out > /sys/class/gpio/gpio68/direction
root@beaglebone:~# echo 0 > /sys/class/gpio/gpio68/value
root@beaglebone:~# echo 1 > /sys/class/gpio/gpio68/value
```

Second setup – the distance sensor is remotized

In this setup, we're going to use BeagleBone Black's serial port to read the measured distance from the ultrasonic sensor.

Using the serial output of the distance sensor

This time, we are interested at the datasheet section where the serial output capability of our sensor is described. In particular, we read:

> *The Pin 5 output delivers asynchronous serial with an RS232 format, except voltages are 0-Vcc. The output is an ASCII capital "R", followed by three ASCII character digits representing the range in centimeters up to a maximum of 765, followed by a carriage return (ASCII 13). The baud rate is 9600, 8 bits, no parity, with one stop bit. Although the voltage of 0-Vcc is outside the RS232 standard, most RS232 devices have sufficient margin to read 0-Vcc serial data. If standard voltage level RS232 is desired, invert, and connect an RS232 converter.*

This is very interesting for two main reasons:

1. The measurement is very precise due to the fact that the sensor gives it to us in a digital format and not by using an analog format (so the measurement is more immune to disturbs.)

2. The information can be sent over a **RS-232** line (even if with some electronic fixes that will be presented soon), which will allow us to have the system core in a different location with respect to the sensor, providing a better usability of the whole system.

So, by using this new setup, the LEDs are still mounted on the BeagleBone Black, while the distance sensor is connected remotely through a RS-232 line. However, we cannot use a classic RS-232 line due to the fact that we still have to supply power to the sensor, and no power can be transferred via a standard RS-232 cable!

The solution is to use a RS-232 connection over a USB cable. In fact, by using a standard USB cable, we are able to send/receive RS-232 data with the needed power supply.

However, some issues are still present:

1. The USB power voltage is 5V, so we need a *USB-to-serial* converter that can manage such voltage level by default, or, is at least 5V tolerant.

2. Reading carefully the preceding snippet of the datasheet, we discover that the output level is TLL and inverted! So, before sending the TX signal to the *USB-to-serial* converter (to the RX pin), we must electrically invert it. (Okay don't panic! I'm going to explain this carefully.)

The solution for the first problem is to use the following *USB-to-serial* converter, which not only works at 3.3V, but also is 5V tolerant.

 The devices can be purchased at the following link (or by surfing the Internet):

`http://www.cosino.io/product/usb-to-serial-converter.`

The datasheet of this device is available at `https://www.silabs.com/Support%20Documents/TechnicalDocs/cp2104.pdf.`

In order to address the second problem, we can use the following circuitry to invert the TTL levels of the TX signal of the sensor:

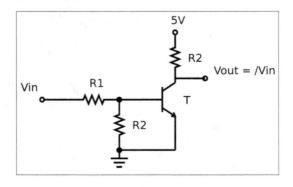

I used the resistors values **R1**=2,2KΩ, **R2**=10KΩ, and a **BC546 transistor** (T). The **Vin** is connected with the sensor's pin 5 (TX), while the **Vout** is connected with the **RX** pin of a **RS232** converter.

The functioning is quite simple—it's a logical NOT port with a voltage level translator. When a logical 0 (a voltage near 0V) is applied to **Vin**, the **transistor** (T) doesn't work, so no current can pass through it and there is no voltage loss on resistor **R2** and the **Vout** is 5V (a logical 1). On the other hand, when a logical 1 (a voltage near 3.3V) is applied to **Vin**, the **transistor** (T) is turned on and a current can now flow through it, and the **Vout** drops down to a voltage near 0V (a logical 0). The following table shows the circuitry functioning in a clear manner, which you can see that it works exactly as we expected!

Vin (V)/logical	Vout (V)/logical
0/0	5/1
3.3/1	0/0

In this situation, the connections to be done on the BeagleBone Black are quite simple. In fact, we have to connect a normal USB cable to the *USB-to-serial* converter and then connect it to the distance sensor, as shown in the following table:

USB-to-serial Pin	Distance sensor pin (label)
GND	7
VBUS	6 (Vcc)
RX	5 (/TX)

 Note that in the table, I used the */TX* electronic notation for the ultrasonic sensor's TX pin (in **C**, we can write *!TX*), since, as already stated, its output signal must be inverted, so, in reality, the TX pin of the distance sensor must be connected with the **Vin** pin of the TTL inverter, while the **Vout** is the effective signal */TX* that must be connected to the USB-to-serial RX pin!

If we decide to use this setup for the distance sensor, the job, from the software point of view, is simpler, since no calibration is needed at all due to the fact that the sensor will return to us the distance in a digital format, that is, without any possible errors due to the analog to digital conversion or voltage scaling, as seen in the preceding section. In fact, we can get the distance simply by reading it from the serial port over the USB connection; so, if everything works well, once we connect the USB cable we should see the following kernel messages:

```
hub 1-0:1.0: state 7 ports 1 chg 0000 evt 0002
usb 1-1: reset full-speed USB device number 2 using musb-hdrc
hub 1-0:1.0: state 7 ports 1 chg 0000 evt 0002
usb 1-1: cp210x converter now attached to ttyUSB0
```

The /dev/ttyUSB0 device is now available:

```
root@beaglebone:~/chapter_02# ls -l /dev/ttyUSB0
crw-rw---T 1 root dialout 188, 0 Apr 23 20:28 /dev/ttyUSB0
```

Now, to read the measurements, we have to configure the serial port as requested by the datasheet with the following command:

```
root@beaglebone:~# stty -F /dev/ttyUSB0 9600
```

Then, the data can be displayed in real-time with the following command:

```
root@beaglebone:~# cat /dev/ttyUSB0
126
```

 You can stop reading by using the *CTRL* + *C* keys.

Connecting the LEDs in the second setup

In this second setup, there is nothing special to say regarding LEDs due to the fact the connections are pretty the same here as in the first setup.

Remember that the LEDs are not related to the USB connection, which is used only to remotize the distance sensor!

The final picture

The following screenshot shows the prototype that I realized to implement this project and to test the software.

Note that I implemented both setups: on the left-half of the breadboard, there is the ultrasonic sensor with related circuitry (that is, the part that can be remotized); on the right-half, there are the circuitry for the LEDs; while in the upper center, there is the inverted voltage translator; and in the lower center, there are the two resistors that implement the voltage divider.

Note also the USB-to-serial converter in the center of the screenshot, where I connected the USB cable that is put into the USB host port of the BeagleBone Black:

I also used an external power supplier due to the fact that the external circuitry and the BeagleBone Black may need more power than what the USB port of your PC can supply!

Setting up the software

In this project, the software is really simple, since we just need a procedure that periodically reads the distance and then turn on and off the LEDs accordingly; however, some issues must be pointed out, especially about how to manage the LEDs and the differences between the two setups of the ultrasonic sensor.

Managing the LEDs

Despite of what was presented in the previous chapter about the GPIO's management, it's important to point out that the Linux kernel has several kinds of devices, each one dedicated to a well-defined usage, and one of these special devices are the led devices, which is a particular type of devices that can be used to manage an LED with different triggers. A **trigger** is a sort of *manager* of the LED that can be programmed to work in a specific manner. Ok, it's better doing an example instead of trying to explain it!

First of all, we have to define the led devices by using a dedicated device tree as reported in the `chapter_02/BB-LEDS-C2-00A0.dts` file in the book's example code repository. The following is a snippet of this file with the relevant code:

```
fragment@1 {
    target = <&ocp>;

    __overlay__ {
      c2_leds {
          compatible     = "gpio-leds";
          pinctrl-names  = "default";
          pinctrl-0      = <&bb_led_pins>;

          white_led {
             label   = "c2:white";
             gpios   = <&gpio3 6 0>;
             linux,default-trigger = "none";
             default-state = "on";
          };

          yellow_led {
             label   = "c2:yellow";
             gpios   = <&gpio3 7 0>;
             linux,default-trigger = "none";
             default-state = "on";
          };
```

```
            red_far_led {
                label   = "c2:red_far";
                gpios   = <&gpio3 2 0>;
                linux,default-trigger = "none";
                default-state = "on";
            };

            red_mid_led {
                label   = "c2:red_mid";
                gpios   = <&gpio3 3 0>;
                linux,default-trigger = "none";
                default-state = "on";
            };

            red_near__led {
                label   = "c2:red_near";
                gpios   = <&gpio3 5 0>;
                linux,default-trigger = "none";
                default-state = "on";
            };
        };
    };
};
```

 Further information regarding how to define Linux's **LED devices** can be found in Linux's source tree in the `linux/Documentation/devicetree/bindings/leds/leds-gpio.txt` file, or online at `https://www.kernel.org/doc/Documentation/devicetree/bindings/leds/leds-gpio.txt`.

As you can see, each GPIO is enabled and defined into the kernel as a LED device by using the `gpio-leds` driver. The code is quite self-explanatory, and it's easy to see that each GPIO definition has a predefined trigger (that is, the default trigger `none`) and the predefined status set to `on`.

To enable this setting, we have to compile it into its binary form by using the `dtc` command as follows:

```
root@beaglebone:~# dtc -O dtb -o /lib/firmware/BB-LEDS-C2-00A0.dtbo -b 0
-@ BB-LEDS-C2-00A0.dts
```

And then, we can load it into the kernel by using the following command:

```
root@beaglebone:~# echo BB-LEDS-C2 > /sys/devices/bone_capemgr.9/slots
```

If everything works well, we should see the following kernel activities:

```
part_number 'BB-LEDS-C2', version 'N/A'
slot #7: generic override
bone: Using override eeprom data at slot 7
slot #7: 'Override Board Name,00A0,Override Manuf,BB-LEDS-C2'
slot #7: Requesting part number/version based 'BB-LEDS-C2-00A0.dtbo'
slot #7: Requesting firmware 'BB-LEDS-C2-00A0.dtbo' for board-name
'Override Board Name', version '00A0'
slot #7: dtbo 'BB-LEDS-C2-00A0.dtbo' loaded; converting to live tree
slot #7: #2 overlays
...
slot #7: Applied #2 overlays.
```

> If we get the following error then we have to disable the **HDMI** support:
>
> ```
> -bash: echo: write error: File exists
> ```
>
> This can be done by editing the uboot settings into the /boot/uboot/uEnv.txt file, and then enabling the following line by uncommenting it:
>
> ```
> optargs=capemgr.disable_partno=BB-BONELT-HDMI,BB-BONELT-HDMIN
> ```
>
> Note that on some BeagleBone Black versions, you may find the uEnv.txt file under the directory /boot instead, and the uboot settings to modify are as follows:
>
> ```
> cape_disable=capemgr.disable_partno=BB-BONELT-HDMI,BB-BONELT-HDMIN
> ```
>
> Then, we only have to reboot the system. Now, if everything was done correctly, we should be able to execute the preceding command without errors.

Note that all the LEDs are now turned on. Now, to manage these new LED devices, we can use the sysfs entries under the following directories:

```
root@beaglebone:~# ls -d /sys/class/leds/c2*
/sys/class/leds/c2:red_far    /sys/class/leds/c2:white
/sys/class/leds/c2:red_mid    /sys/class/leds/c2:yellow
/sys/class/leds/c2:red_near
```

As you can see, all the names we used in the DTS file are present, and we also find the following files in each directory:

```
root@beaglebone:~# ls /sys/class/leds/c2\:white
brightness  device  max_brightness  power  subsystem  trigger  uevent
```

The relevant files are `trigger`, `brightness`, and `max_brightness`. The `trigger` file is used to discover the current trigger, and, in case, to change it. In fact, by reading the file, we see the following:

```
root@beaglebone:~# cat /sys/class/leds/c2\:white/trigger
[none] nand-disk mmc0 mmc1 timer oneshot heartbeat backlight gpio cpu0
default-on transient
```

As we expected, the current trigger is `none` (the one between the square brackets), and we can change it simply by writing the new name into the same file (see the preceding example).

The `brightness` and `max_brightness` files are specific of the current trigger and can be used to set the brightness of the LED from the `0` value to the maximum value stored in the `max_brightness` file. Just to test it, we can read the current values into these files to verify that the current status is at the maximum brightness:

```
root@beaglebone:~# cat /sys/class/leds/c2\:white/max_brightness
255
root@beaglebone:~# cat /sys/class/leds/c2\:white/brightness
255
```

To turn off the LED, we can use the following command:

```
root@beaglebone:~/# echo 0 > /sys/class/leds/c2\:white/brightness
```

 Note that our LEDs are two functional values only, that is, `0` and `255`, due to the fact that the LEDs we are using have only two valid states.

However, having a flashing capability in our project for when the car is near a specific distance from the wall in such a way that gives a better warning about the increasing *danger* could be very interesting. In particular, we can do this in such a way that when the red LEDs must be turned on, according to what was stated in the *Connecting the LEDs in the first setup* section, in this chapter, the frequency of the flash will keep increasing as the distance reduces, they will stop flashing and remain turned on when the distance is less than 0.10 m.

To flash an LED with the desired frequency, we can use the `timer` trigger. In order to show how it works, let's try to enable it on the LED named `red_far` by using the following command:

```
root@beaglebone:~# echo timer > /sys/class/leds/c2\:red_far/trigger
```

After executing this command, the LED should start flashing; then looking again into the directory, we see that new files are now available:

```
root@beaglebone:~# ls /sys/class/leds/c2\:red_far
brightness   delay_on    max_brightness   subsystem   uevent
delay_off    device      power            trigger
```

The new interesting files are `delay_on` and `delay_off`, and they can be used to define instead how long the LED must be turned on and how long it must be turned off. It's quite obvious that the LED's blinking frequencies (F) can now be set with the following formula:

$F = 1 / T$, where $T = T_{delay\_on} + T_{delay\_off}$

So, for instance, if we wish that the LED will flash with a frequency of 10Hz, we can use the following commands:

```
root@beaglebone:~# echo 50 > /sys/class/leds/c2\:red_far/delay_on
root@beaglebone:~# echo 50 > /sys/class/leds/c2\:red_far/delay_off
```

The value 50 means: 50ms at *on* state and 50ms at *off* state. So, we have T_{delay_on}=50ms and T_{delay_off}=50ms, so T=100ms, and then F=10Hz.

Considering that the human eye is still sensitive at more or less 25Hz at maximum, and that the minimum allowed frequency is 1Hz, the possible values to be written into each of the preceding two files are from 500 (ms) for a blinking frequency of 1Hz to 20 (ms) for a blinking frequency of 25Hz.

A possible implementation of a controlling program for the LEDs is in the `chapter_02/led_set.sh` file in the book's example code repository. The following code is a snippet of the relevant code:

```
case $mode in
-1)
    # Turn on the LED
    echo none > /sys/class/leds/c2\:$name/trigger
    echo 255 > /sys/class/leds/c2\:$name/brightness
    ;;
```

```
0)
   # Turn off the LED
   echo none > /sys/class/leds/c2\:$name/trigger
   echo 0 > /sys/class/leds/c2\:$name/brightness
   ;;

*)
   # Flash the LED
   t=$((1000 / $mode / 2))

   echo timer > /sys/class/leds/c2\:$name/trigger
   echo $t > /sys/class/leds/c2\:$name/delay_on
   echo $t > /sys/class/leds/c2\:$name/delay_off
   ;;
esac
```

Here, the code turns on the LED addressed by the name variable in case the mode variable is set to -1, while it turns it off the same LED when mode is set to 0. Also, the code will enable the timer trigger with proper settings in case the mode variable is between the values 1 and 25(Hz).

The following is a sample usage:

```
root@beaglebone:~# ./led_set.sh red_far -1
root@beaglebone:~# ./led_set.sh red_far 0
root@beaglebone:~# ./led_set.sh red_far 10
```

The distance monitor

Now it's time to see how our park assistant can work in practice. A possible implementation of the code is reported in the chapter_02/distance_mon.sh script in the book's example code repository. The following code snippet shows the main code:

```
# Ok, do the job
while sleep .1 ; do
   # Read the current distance from the sensor
   d=$($d_fun)
   dbg "d=$d"

   # Manage the LEDs
   leds_man $d
done
```

The functioning is simple — the code periodically reads the distance from the sensor by using the function pointed by the d_fun variable, and then turns the LEDs on and off, according to the value of the distance d (in cm) by using the leds_man function.

The d_fun variable holds the name of the function that should read the distance by using the ADC, that is, read_adc, or the name of the function that uses the serial port, that is, read_tty. The following are the two functions:

```
function read_adc () {
    n=$(cat $ADC_DEV)

    d=$(bc -l <<< "$k * 3.3 * $n/4095 / 0.00161")
    printf "%.0f\n" $d
}

function read_tty () {
    while read d < $TTY_DEV ; do
        [[ "$d" =~ R[0-9]{2,3} ]] && break
    done

    # Drop the "R" character
    d=${d#R}

    # Drop the leading "0"
    echo ${d#0}
}
```

Note that the read_adc file uses the bc program to calculate the translation formula discussed before, while the read_tty uses the Bash's read and while commands to read a complete data line (which is in the form Rxxx\r, as reported in the datasheet.)

 The bc command may be not installed by default into the BeagleBone Black's distribution, so you can install it by using the following command:

root@beaglebone:~# aptitude install bc

The leds_man function is as follows:

```
function leds_man () {
    d=$1

    # Calculate the blinking frequency with the following
    # fixed values:
    #     f=1Hz  if d=100cm
    #     f=25Hz if d=25cm
```

```
f=$((25 - 21 * ( d - 25 ) / 75))
[ $f -gt 25 ] && f=25
[ $f -lt 1 ] && f=1

if [ "$d" -gt 500 ] ; then
    ./led_set.sh white       0
    ./led_set.sh yellow      0
    ./led_set.sh red_far     0
    ./led_set.sh red_mid     0
    ./led_set.sh red_near    0

    return
fi

if [ "$d" -le 500 -a "$d" -gt 200 ] ; then
    ./led_set.sh white      -1
    ./led_set.sh yellow      0
    ./led_set.sh red_far     0
    ./led_set.sh red_mid     0
    ./led_set.sh red_near    0

    return
fi

if [ "$d" -le 200 -a "$d" -gt 100 ] ; then
    ./led_set.sh white      -1
    ./led_set.sh yellow     -1
    ./led_set.sh red_far     0
    ./led_set.sh red_mid     0
    ./led_set.sh red_near    0

    return
fi

if [ "$d" -le 100 -a "$d" -gt 50 ] ; then
    ./led_set.sh white      -1
    ./led_set.sh yellow     -1
    ./led_set.sh red_far    $f
    ./led_set.sh red_mid     0
    ./led_set.sh red_near    0

    return
fi
```

```
    if [ "$d" -le 50 -a "$d" -gt 20 ] ; then
        ./led_set.sh white     -1
        ./led_set.sh yellow    -1
        ./led_set.sh red_far   -1
        ./led_set.sh red_mid   $f
        ./led_set.sh red_near   0

        return
    fi

    # if -le 20
    ./led_set.sh white     -1
    ./led_set.sh yellow    -1
    ./led_set.sh red_far   -1
    ./led_set.sh red_mid   -1
    ./led_set.sh red_near  -1
}
```

The function first calculates the blinking frequency in order to respect what was stated in the preceding sections, and then it uses a big case to decide which LEDs configuration must be used to notify the driver.

Final test

To test the prototype, we must first select one setup and perform the needed connections, as stated before. Then we have to turn on the board.

After the login, we must setup the system by using the commands discussed before, or simply by using the chapter_02/SYSINIT.sh command in the book's example code repository. Then, we must execute the distance_mon.sh command accordingly.

Note that looking into the SYSINIT.sh file, you can read:

```
# Uncomment the following in case of buggy kernel in
USB host management
# cat /dev/bus/usb/001/001 > /dev/null ; sleep 1
```

This is if after plugging in the USB cable, you get an error in recognizing the /dev/ttyUSB0 device.

To test my prototype using the first setup, I used the following command:

```
root@beaglebone:~# ./distance_mon.sh -d -k 1.38 adc
distance_mon.sh: d_fun=read_adc k=1.38
distance_mon.sh: d=176
distance_mon.sh: d=175
distance_mon.sh: d=175
distance_mon.sh: d=175
distance_mon.sh: d=175

...
```

On the other hand, to test the second one, I used this other command:

```
root@beaglebone:~# ./distance_mon.sh -d serial
distance_mon.sh: d_fun=read_tty k=1
distance_mon.sh: d=151
distance_mon.sh: d=152
distance_mon.sh: d=151
distance_mon.sh: d=152
distance_mon.sh: d=152

...
```

You can stop the program by using the *CTRL + C* keys.

Summary

In this chapter, we discovered how to manage an ultrasonic sensor in two different manners, by using an ADC and by a serial connection over a USB cable, in order to have two different setups of the same device: one with all peripherals on the BeagleBone Black and one where a sensor is remotized by using a USB connection.

Also, we learned how to manage Linux's LED devices that allow us to have different usage of a simple GPIO line by kernel features.

In the next chapter, we'll see how to realize an aquarium monitor in which we'll be able to record all the environment data, and then we'll see how to control the life of our be loved fishes from a web panel.

3
Aquarium Monitor

In this chapter, we'll see how to realize an aquarium monitor where we'll be able to record all the environment data and then control the life of our loved fish from a web panel.

By using specific sensors, you'll learn how to monitor your aquarium with the possibility to set alarms, log the aquarium data (water temperature), and to perform actions such as cooling the water and feeding the fish.

Simply speaking, we're going to implement a simple aquarium web monitor with a real-time live video, some alarms in case of malfunctioning, and a simple temperature data logging that allows us to monitor the system from a standard PC as well as from a smartphone or tablet, without using any specifying mobile app, but just using the on-board standard browser only.

The basics of functioning

This aquarium monitor is a good (even if very simple) example about how a web monitoring system should be implemented, giving to the reader some basic ideas about how a mid-complex system works and how we can interact with it in order to modify some system settings, displaying some alarms in case of malfunctioning, and plotting a data logging on a PC, smartphone, or tablet.

Despite these aspects, the basic functioning of this project is similar to what we've already done in previous chapters: we have a periodic task that collects the data and then decides what to do. However, this time, we have a user interface (the web panel) to manage, and a video streaming to be redirected into a web page.

Note also that in this project, we need an additional power supply in order to power and manage 12V devices (such as a water pump, a lamp, and a cooler) with the BeagleBone Black, which is powered at 5V instead.

 Note that I'm not going to test this prototype on a real aquarium (since I don't have one), but by using a normal tea cup filled with water! So you should consider this project for educational purpose only, even if, with some enhancements, it could be used on a real aquarium too!

Setting up the hardware

About the hardware, there are at least two major issues to be pointed out:

- **Power supply**: We have two different voltages to use due to the fact the water pump, the lamp, and the cooler are 12V powered, while the other devices are 5V/3.3V powered. So, we have to use a dual output power source (or two different power sources) to power up our prototype.

- **Interface**: The second issue is about using a proper interface circuitry between the 12V world and the 5V one in such a way that it doesn't damage the BeagleBone Black or other devices. Let me point out that a single GPIO of the BeagleBone Black can manage a voltage of 3.3V, so we need a proper circuitry to manage a 12V device.

Setting up the 12V devices

As just stated, these devices need special attention and a dedicated 12V power line which, of course, cannot be the one we wish to use to supply the BeagleBone Black. On my prototype, I used a 12V power supplier that can supply a current till 1A. These characteristics should be enough to manage a single water pump, a lamp, and a cooler.

After you get a proper power supplier, we can pass to show the circuitry to use to manage the 12V devices. Since all of them are simple on/off devices, we can use a relay to control them. I used the device shown in the following image where we have 8 relays:

 The devices can be purchased at the following link (or by surfing the Internet): http://www.cosino.io/product/5v-relays-array

Then, the schematic to connect a single 12V device is shown in the following diagram:

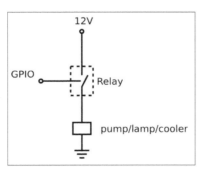

Simply speaking, for each device, we can turn the power supply on and off simply by moving a specific GPIO of our BeagleBone Black. Note that each relays of the array board can be managed in direct or inverse logic by simply choosing the right connections accordingly as reported on the board itself, that is, we can decide that, by putting the GPIO into a logic 0 state, we can activate the relay, and then, turning on the attached device, while putting the GPIO into a logic 1 state, we can deactivate the relay, and then turn off the attached device.

By using the following logic, when the LED of a relay is turned on, the corresponding device is powered on.

The BeagleBone Black's GPIOs and the pins of the relays array I used with 12V devices are reported in the following table:

Pin	Relays Array pin	12V Device
P8.10 - GPIO66	3	Lamp
P8.9 - GPIO69	2	Cooler
P8.12 - GPIO68	1	Pump
P9.1 - GND	GND	
P9.5 - 5V	Vcc	

To test the functionality of each GPIO line, we can use the following command to set up the GPIO as an output line at high state:

```
root@arm:~# ./bin/gpio_set.sh 68 out 1
```

 Note that the *off* state of the relay is 1, while the *on* state is 0.

Then, we can turn the relay on and off by just writing 0 and 1 into /sys/class/gpio/gpio68/value file, as follows:

```
root@arm:~# echo 0 > /sys/class/gpio/gpio68/value
root@arm:~# echo 1 > /sys/class/gpio/gpio68/value
```

Setting up the webcam

The webcam I'm using in my prototype is a normal UVC-based webcam, but you can safely use another one that is supported by the **mjpg-streamer** tool.

 See the mjpg-streamer project's home site for further information at http://sourceforge.net/projects/mjpg-streamer/.

Once connected to the BeagleBone Black USB host port, I get the following kernel activities:

```
usb 1-1: New USB device found, idVendor=045e, idProduct=0766
usb 1-1: New USB device strings: Mfr=1, Product=2, SerialNumber=0
```

```
usb 1-1: Product: Microsoft LifeCam VX-800
usb 1-1: Manufacturer: Microsoft
...
uvcvideo 1-1:1.0: usb_probe_interface
uvcvideo 1-1:1.0: usb_probe_interface - got id
uvcvideo: Found UVC 1.00 device Microsoft LifeCam VX-800 (045e:0766)
```

Now, a new driver called `uvcvideo` is loaded into the kernel:

```
root@beaglebone:~# lsmod
Module                Size  Used by
snd_usb_audio        95766  0
snd_hwdep             4818  1 snd_usb_audio
snd_usbmidi_lib      14457  1 snd_usb_audio
uvcvideo             53354  0
videobuf2_vmalloc     2418  1 uvcvideo
...
```

Okay, now, to have a streaming server, we need to download the mjpg-streamer source code and compile it. We can do everything within the BeagleBone Black itself with the following command:

```
root@beaglebone:~# svn checkout svn://svn.code.sf.net/p/mjpg-streamer/
code/ mjpg-streamer-code
```

> The svn command is part of the `subversion` package and can be installed by using the following command:
> `root@beaglebone:~# aptitude install subversion`

After the download is finished, we can compile and install the code by using the following command line:

```
root@beaglebone:~# cd mjpg-streamer-code/mjpg-streamer/ && make && make
install
```

> You can find a compressed archive copy of this program in the `chapter_03/mjpg-streamer-code.tgz` file in the book's example code repository.

If no errors are reported, you should now be able to execute the new command as follows, where we ask for the help message:

```
root@beaglebone:~# mjpg_streamer --help
-------------------------------------------------------------------
Usage: mjpg_streamer
  -i | --input "<input-plugin.so> [parameters]"
  -o | --output "<output-plugin.so> [parameters]"
  [-h | --help ]........: display this help
  [-v | --version ].....: display version information
  [-b | --background]...: fork to the background, daemon mode
...
```

> If you get an error like the following it means that your system misses the convert tool:
>
> ```
> make[1]: Entering directory `/root/mjpg-streamer-
> code/mjpg-streamer/plugins/input_testpicture'
> convert pictures/960x720_1.jpg -resize 640x480!
> pictures/640x480_1.jpg
> /bin/sh: 1: convert: not found
> make[1]: *** [pictures/640x480_1.jpg] Error 127
> ```
>
> You can install it by using the usual aptitude command:
>
> ```
> root@beaglebone:~# aptitude install imagemagick
> ```

Okay, now we are ready to test the webcam. Just run the following command line and then point a web browser to the address `http://192.168.32.46:8080/?action=stream` (where you should replace my IP address `192.168.32.46` with your BeagleBone Black's one) in order to get the live video from your webcam:

```
root@beaglebone:~# LD_LIBRARY_PATH=/usr/local/lib/ mjpg_streamer -i
"input_uvc.so -y -f 10 -r QVGA" -o "output_http.so -w /var/www/"
```

> Note that you can use the USB Ethernet address `192.168.7.2` too if you're not using the BeagleBone Black's Ethernet port.

If everything works well, you should get something similar to what is shown in the following screenshot:

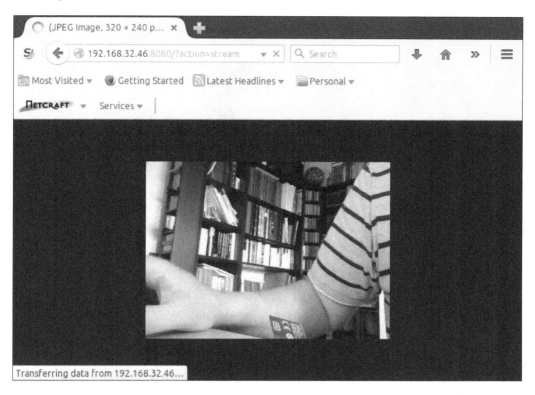

> If you get an error, as follows it means that some other process is holding the `8080` port, and most probably it's occupied by the `Bone101` service:
>
> **bind: Address already in use**
>
> To disable it, you can use the following commands:
>
> `root@BeagleBone:~# systemctl stop bonescript.socket`
> `root@BeagleBone:~# systemctl disable bonescript.socket`
> `rm '/etc/systemd/system/sockets.target.wants/`
> `bonescript.socket'`
>
> Or, you can simply use another port, maybe port `8090`, with the following command line:
>
> `root@beaglebone:~# LD_LIBRARY_PATH=/usr/local/lib/`
> `mjpg_streamer -i "input_uvc.so -y -f 10 -r QVGA" -o`
> `"output_http.so -p 8090 -w /var/www/"`

Connecting the temperature sensor

The temperature sensor used in my prototype is the one shown in the following image:

 The devices can be purchased at the following link (or by surfing the Internet): http://www.cosino.io/product/waterproof-temperature-sensor.

The datasheet of this device is available at http://datasheets.maximintegrated.com/en/ds/DS18B20.pdf.

As you can see, it's a waterproof device, so we can safely put it into the water to get its temperature.

This device is a **1-Wire** device and we can get access to it by using the w1-gpio driver, which emulates a 1-Wire controller by using a standard BeagleBone Black GPIO pin. The electrical connection must be done according to the following table, keeping in mind that the sensor has three colored connection cables:

Pin	Cable color
P9.4 - Vcc	Red
P8.11 - GPIO1_13	White
P9.2 - GND	Black

 Interested readers can follow this URL for more information about how 1-Wire works: http://en.wikipedia.org/wiki/1-Wire

Keep in mind that, since our 1-Wire controller is implemented in software, we have to add a pull-up resistor of 4.7KΩ between the red and white cable in order to make it work!

Once all connections are in place, we can use the `chapter_03/BB-W1-GPIO-00A0.dts` file in the book's example code repository to enable the 1-Wire controller on the *P8.11* pin of the BeagleBone Black's expansion connector. The following snippet shows the relevant code where we enable the `w1-gpio` driver and assign to it the proper GPIO:

```
fragment@1 {
    target = <&ocp>;

    __overlay__ {
        #address-cells  = <1>;
        #size-cell      = <0>;
        status          = "okay";

        /* Setup the pins */
        pinctrl-names   = "default";
        pinctrl-0       = <&bb_w1_pins>;

        /* Define the new one-wire master as based on w1-gpio
        * and using GPIO1_13
        */
        onewire@0 {
            compatible      = "w1-gpio";
            gpios           = <&gpio2 13 0>;
        };
    };
};
```

To enable it, we must use the `dtc` program to compile it as follows:

```
root@beaglebone:~# dtc -O dtb -o /lib/firmware/BB-W1-GPIO-00A0.dtbo -b 0
-@ BB-W1-GPIO-00A0.dts
```

Then, we have to load it into the kernel with the following command:

```
root@beaglebone:~# echo BB-W1-GPIO > /sys/devices/bone_capemgr.9/slots
```

If everything works well, we should see a new 1-Wire device under the `/sys/bus/w1/devices/` directory, as follows:

```
root@beaglebone:~# ls /sys/bus/w1/devices/
28-000004b541e9  w1_bus_master1
```

The new temperature sensor is represented by the directory named
28-000004b541e9. To read the current temperature, we can use the
cat command on the w1_slave file as follows:

```
root@beaglebone:~# cat /sys/bus/w1/devices/28-000004b541e9/w1_slave
d8 01 00 04 1f ff 08 10 1c : crc=1c YES
d8 01 00 04 1f ff 08 10 1c t=29500
```

 Note that your sensors have a different ID, so in your system you'll
get a different path name in the /sys/bus/w1/devices/28-
NNNNNNNNNNNN/w1_slave form.

In the preceding example, the current temperature is t=29500, which is expressed in
millicelsius degree (m°C), so it's equivalent to 29.5°C.

 The reader can take a look at the book *BeagleBone Essentials*, *Packt
Publishing*, written by the author of this book, in order to have more
information regarding the management of the 1-Wire devices on the
BeagleBone Black.

Connecting the feeder

The fish feeder is a device that can release some feed by moving a motor. Its
functioning is represented in the following screenshot:

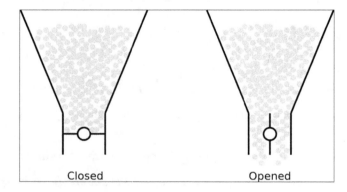

Closed   Opened

In the closed position, the motor is at horizontal position, so the feed cannot fall down, while in the **Open** position, the motor is at vertical position, so that the feed can fall down. I have no real fish feeder, but looking at the preceding functioning, we can simulate it by using the servo motor shown in the following screenshot:

The device can be purchased at the following link (or by surfing the Internet): http://www.cosino.io/product/nano-servo-motor.

The datasheet of this device is available at http://hitecrcd.com/files/Servomanual.pdf.

This device can be controlled in position, and it can rotate by 90 degrees with a proper **PWM** signal in input. In fact, reading into the datasheet, we discover that the servo can be managed by using a periodic square waveform with a **period** (T) of 20ms and with a **high state time** (t_{high}) between 0.9ms and 2.1ms with 1.5ms as (more or less) center. So, we can consider the motor in the **Open** position when $t_{high}=1ms$ and in the **Close** position when $t_{high}=2ms$ (of course, these values should be carefully calibrated once the feeder is really built up!).

Let's connect the servo as described by the following table:

Pin	Cable color
P9.3 - Vcc	Red
P9.22 - PWM	Yellow
P9.1 - GND	Black

[Interested readers can find more details about the PWM at https:// en.wikipedia.org/wiki/Pulse-width_modulation.]

To test the connections, we have to enable one PWM generator of the BeagleBone Black. So, to respect the preceding connections, we need the one which has its output line on pin P9.22 of the expansion connectors. To do it, we can use the following commands:

```
root@beaglebone:~# echo am33xx_pwm > /sys/devices/bone_capemgr.9/slots
root@beaglebone:~# echo bone_pwm_P9_22 > /sys/devices/bone_capemgr.9/
slots
```

Then, in the /sys/devices/ocp.3 directory, we should find a new entry related to the new enabled PWM device, as follows:

```
root@beaglebone:~# ls -d /sys/devices/ocp.3/pwm_*
/sys/devices/ocp.3/pwm_test_P9_22.12
```

Looking at the /sys/devices/ocp.3/pwm_test_P9_22.12 directory, we see the files we can use to manage our new PWM device:

```
root@beaglebone:~# ls /sys/devices/ocp.3/pwm_test_P9_22.12/
driver duty    modalias   period   polarity   power    run     subsystem
uevent
```

As we can deduce from the preceding file names, we have to properly set up the values into the files named as polarity, period, and duty. So, for instance, the center position of the servo can be achieved by using the following commands:

```
root@beaglebone:~# echo 0 > /sys/devices/ocp.3/pwm_test_P9_22.12/polarity
root@beaglebone:~# echo 20000000 > /sys/devices/ocp.3/pwm_test_P9_22.12/
period
root@beaglebone:~# echo 1500000 > /sys/devices/ocp.3/pwm_test_P9_22.12/
duty
```

The polarity is set to 0 to invert it, while the values written in the other files are time values expressed in nanoseconds, set at a period of 20ms and a duty cycle of 1.5ms, as requested by the datasheet (time values are all in nanoseconds.) Now, to move the gear totally clockwise, we can use the following command:

```
root@beaglebone:~# echo 2100000 > /sys/devices/ocp.3/pwm_test_P9_22.12/
duty
```

On the other hand, the following command is to move it totally anticlockwise:

```
root@beaglebone:~# echo 900000 > /sys/devices/ocp.3/pwm_test_P9_22.12/
duty
```

So, by using the following command sequence, we can open and then close (with a delay of 1 second) the gate of the feeder:

```
echo 1000000 > /sys/devices/ocp.3/pwm_test_P9_22.12/duty
sleep 1
echo 2000000 > /sys/devices/ocp.3/pwm_test_P9_22.12/duty
```

Note that by simply modifying the delay, we can control how much feed should fall down when the feeder is activated.

 The script that implements the feeder controlling mechanism can be found in the chapter_03/feeder.sh file in the book's example code repository.

The water sensor

The water sensor I used is shown in the following screenshot:

 The device can be purchased at the following link (or by surfing the Internet): http://www.cosino.io/product/water_sensor.

This is a really simple device that implements what is shown in the following screenshot, where the **resistor (R)** has been added to limit the current when the water closes the circuit:

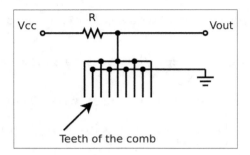

When a single drop of water *touches* two or more teeth of the comb in the schematic, the circuit is closed and the **output voltage (Vout)** drops from **Vcc** to 0V.

So, if we wish to check the water level in our aquarium, that is, if we wish to check for a water leakage, we can imagine to put the aquarium into some sort of saucer, and then this device into it, so, if a water leakage occurs, the water is collected by the saucer, and the output voltage from the sensor should move from **Vcc** to GND.

The GPIO used for this device are shown in the following table:

Pin	Cable color
P9.3 - 3.3V	Red
P8.16 - GPIO67	Yellow
P9.1 - GND	Black

To test the connections, we have to define GPIO 67 as an input line with the following command:

```
root@beaglebone:~# ../bin/gpio_set.sh 67 in
```

Then, we can try to read the GPIO status while the sensor is in the water and when it is not, by using the following two commands:

```
root@beaglebone:~# cat /sys/class/gpio/gpio67/value
0
root@beaglebone:~# cat /sys/class/gpio/gpio67/value
1
```

The final picture

The following screenshot shows the prototype I realized to implement this project and to test the software. As you can see, the aquarium has been replaced by a cup of water:

Note that we have two external power suppliers: the usual one at 5V for the BeagleBone Black, and the other one with an output voltage of 12V for the other devices (you can see its connector in the upper-right corner on the right of the webcam.)

Setting up the software

Regarding the software, this time the major part is covered by the web interface, which is the real core of the project, and the acquisition and controlling process to get the aquarium data and managing the actuators. Then, a dedicated monitor script will be used to implement the communication between the web interface and the internal database.

Managing the system status and configuration

To manage the status of all devices and to do the data logging, we can use a database again to store all the relevant data, as we did in *Chapter 1, Dangerous Gas Sensors*. So, we can use the chapter_03/my_init.sh file in the book's example code repository to set up the database. The following snippet shows the relevant code where we define the tables used in the project:

```
# Select database
USE aquarium_mon;

#
# Create the system status table
#

CREATE TABLE status (
        n VARCHAR(64) NOT NULL,
        v VARCHAR(64) NOT NULL,
        PRIMARY KEY (n)
) ENGINE=MEMORY;

# Setup default values
INSERT INTO status (n, v) VALUES('alarm_sys', '0');
INSERT INTO status (n, v) VALUES('alarm_level', '0');
INSERT INTO status (n, v) VALUES('alarm_temp', '0');
INSERT INTO status (n, v) VALUES('water', '21');
INSERT INTO status (n, v) VALUES('cooler', '0');
INSERT INTO status (n, v) VALUES('pump', '0');
INSERT INTO status (n, v) VALUES('lamp', '0');
INSERT INTO status (n, v) VALUES('force_cooler', '0');
INSERT INTO status (n, v) VALUES('force_pump', '0');
INSERT INTO status (n, v) VALUES('force_lamp', '0');
INSERT INTO status (n, v) VALUES('force_feeder', '0');

#
# Create the system configuration table
#

CREATE TABLE config (
    n VARCHAR(64) NOT NULL,
    v VARCHAR(64) NOT NULL,
    PRIMARY KEY (n)
);
```

```
# Setup default values
INSERT INTO config (n, v) VALUES('pump_t_on', '20');
INSERT INTO config (n, v) VALUES('pump_t_off', '60');
INSERT INTO config (n, v) VALUES('feeder_interval', '60');
INSERT INTO config (n, v) VALUES('water_temp_max', '27');
INSERT INTO config (n, v) VALUES('water_temp_min_alarm', '18');
INSERT INTO config (n, v) VALUES('water_temp_max_alarm', '29');

#
# Create one table per sensor data
#

CREATE TABLE temp_log (
    t DATETIME NOT NULL,
    v FLOAT,
    PRIMARY KEY (t)
);
```

The `status` table holds the system status variables with the following meanings:

Variable name	Description
alarm_sys	Generic system alarm (I/O and communication errors, and so on).
alarm_level	A water leakage has been detected.
alarm_temp	Water temperature is over the `water_temp_max_alarm` value or under the `water_temp_min_alarm` value (in °C).
Water	Current water temperature (in °C).
Cooler	Current cooler status (0 = off, 1 = on).
Pump	Current pump status (0 = off, 1 = on).
Lamp	Current lamp status (0 = off, 1 = on).
force_cooler	The user asks to turn on the cooler.
force_pump	The user asks to turn on the pump.
force_lamp	The user asks to turn on the lamp.
force_feeder	The user asks to enable the feeder.

Note that the feeder has no current status variable due to the fact that it cannot stay opened or closed, but just has an *impulse-on* functioning; that is, when enabled, it opens and then closes it's internal gate.

On the other hand, in the `config` table, there are system configuration variables with the following meanings:

Variable name	Description
pump_t_on	Time in seconds when the pump must be on.
pump_t_off	Time in seconds when the pump must be off.
feeder_interval	Time in seconds between two consecutive "lunch time".
water_temp_max	Turn the cooler on if the water temperature is over this value (in °C).
water_temp_min_ alarm	Turn the water alarm on if the water temperature is under this value (in °C).
water_temp_max_ alarm	Turn the water alarm on if the water temperature is over this value (in °C).

Note that my project misses the `water_temp_min` configuration variable due to the fact that I have no heater available to increase the water temperature in case it is too cold. However, the reader should have all needed information to fill this gap by reading this chapter!

In the end, the `temp_log` table is used to store all water temperature measurements useful to show a little graph into the user's control panel (see the following section).

Building up the web control panel

The web control panel is written in **PHP** and **JavaScript**. PHP is used to implement the acquisition and controlling processes plus the main page, while the JavaScript is used to implement the graphical widgets. In particular, this last part has been realized by using an interesting toolkit named **Drinks** (http://www.goincompany. com/drinks.php).

Using the widgets implemented by this toolkit is trivial. In order to install it, we just need to download the zip archive from the project's home site, unzip it, and then move the files with the .js extension into the web server's root directory. On my BeagleBone Black, I'm running the Apache web server, which has its root directory in the /var/www directory. So, to install the Drinks toolkit, I moved the files as follows:

```
root@beaglebone:~/chapter_03# ls /var/www/

Display.js  Knob.js  Slider.js  Drinks.js  Led.js  Switch.js
```

Now, we have to add the code to build up our web control panel and manage it in this directory. The main script can be found in the `chapter_03/aquarium.php` file in the book's example code repository. I'm going to report all its relevant code into several snippets now.

In the following first snippet, there is the PHP code to get the initial statuses of the input widgets, that is, those widgets that the user directly manages to send commands to the system. During the first loading of this page, the user will get the current status of this widgets as they are stored into the internal database:

```
# Open the DB
db_open("aquarium_mon");

# Set initial statuses for input widgets
$force_cooler = db_get_status("force_cooler");
$force_pump = db_get_status("force_pump");
$force_lamp = db_get_status("force_lamp");
$force_feeder = db_get_status("force_feeder");
```

Then follows the header of the **HTML** page as follows:

```html
<html>
   <head>
      <link href="aquarium.css" rel="stylesheet" type="text/css">

      <script type="text/javascript" src="Drinks.js"></script>

      <script>
         var man_in = Drinks.createManager();
         man_in.href = 'handler.php';
         man_in.input = new Array("force_cooler", "force_pump",
            "force_lamp", "force_feeder");
         man_in.refresh = 1;
         man_in.start();

         var man_out = Drinks.createManager();
         man_out.href = 'handler.php';
         man_out.refresh = 1;
         man_out.start();
      </script>
   </head>
```

By using this code, we instruct the Drinks toolkit to generate two managers—one to manage the input widgets (man_in), and the other to manage all other output widgets (man_out). The output widgets, as opposed to the input ones, are all the widgets that are not under the direct control of the user and then are updated by the system to show the system status to the user.

Both managers will refresh their internal status each second (refresh=1), and both will use the external handler named handler.php to do it. The code of this handler (which will be presented in the next section) are executed periodically, and it's used to get the input widget's statuses and to set the output widgets statuses into the control panel.

Then, the control panel is divided into three main subsections. The first one is where the live video and the alarms are placed. This can be achieved with the following code snippet:

```
<table>
    <tr>
        <th><h3>Live video</h3></th>
        <th><h3>Alarms</h3></th>
    </tr>
    <tr>
        <td>
            <img src="http://<?=$_SERVER["SERVER_
ADDR"]?>:8080/?action=stream" alt="real-time video feed" />
        </td>
        <td>
            <table class="widget">
                <tr>
                    <th>system</th>
                    <th>Water level</th>
                    <th>Water temperature</th>
                </tr>
                <tr>
                    <td><led id="alarm_sys" type="round" radius="25"
color="red"></led></td>
                    <td><led id="alarm_level" type="round" radius="25"
color="red"></led></td>
                    <td><led id="alarm_temp" type="round" radius="25"
color="red"></led></td>
                </tr>
            </table>
        </td>
    </tr>
</table>
```

Here, the following line is used to enable the live video from the webcam:

```
<img src="http://<?=$_SERVER["SERVER_ADDR"]?>:8080/?action=stream"
alt="real-time video feed" />
```

In the end, the following three lines are used to define the alarm LEDs related to the corresponding alarm variables:

```
                <td><led id="alarm_sys" type="round" radius="25"
color="red"></led></td>
                <td><led id="alarm_level" type="round" radius="25"
color="red"></led></td>
                <td><led id="alarm_temp" type="round" radius="25"
color="red"></led></td>
```

The second subsection holds the control widgets, that is, the water thermometer and the lamp, cooler, pump, and feeder LEDs and switches. Since all the input widgets are here, the code uses a big HTML form where these items are placed:

```
    <form method="post">
        <table class="widget">
            <tr>
                <th>Water temp (C)</th>
                <th>Lamp</th>
                <th>Cooler</th>
                <th>Pump</th>
                <th>Feeder</th>
            </tr>
            <tr>
                <td>
                    <display id="water" type="thermo" max_range="50" range_
from="10" range_to="50" autoscale="true"></display>
                </td>
                <td>
                    <led id="lamp" type="round" radius="25"></led>
                    <switch id="force_lamp" type="circle" value="<?=$force_
lamp?>"></switch>
                </td>
                <td>
                    <led id="cooler" type="round" radius="25"></led>
                    <switch id="force_cooler" type="circle"
value="<?=$force_cooler?>"></switch>
                </td>
                <td>
                    <led id="pump" type="round" radius="25"></led>
                    <switch id="force_pump" type="circle" value="<?=$force_
pump?>"></switch>
                </td>
```

```
          <td>
            <led id="feeder" type="round" radius="25"></led>
            <switch id="force_feeder" type="toggle" width="80"
value="<?=$force_feeder?>"></switch>
          </td>
        </tr>
      </table>
      <input type="hidden">
    </form>
```

The following line is used to display a thermometer that reports the water temperature:

```
<display id="water" type="thermo" max_range="50" range_from="10"
range_to="50" autoscale="true"></display>
```

The following two lines are used to display the lamp, cooler, and pump switches with the relative LED that indicates the device status:

```
            <led id="lamp" type="round" radius="25"></led>
            <switch id="force_lamp" type="circle" value="<?=$force_
lamp?>"></switch>
```

When an LED is turned on, the corresponding device is turned on, while when it is turned off, the device is turned off. On the other hand, by moving one of the preceding switches, the user can force the system to turn on the corresponding device during the *next cycle*. (I'm going to explain what the *next cycle* term means in the next section.)

A special notice must be applied for the feeder. As already stated, it can be enabled with an impulse, and not simply turned on or off. To highlight this fact, this time I used a different kind of switch widget. So, the LED is used to notify the user that the feeder will be enabled in the next cycle, while the LED will remain on and it will be turned off once the feeder has been really enabled.

The code to display the feeder controls is shown in the following code snippet:

```
            <led id="feeder" type="round" radius="25"></led>
            <switch id="force_feeder" type="toggle" width="80"
value="<?=$force_feeder?>"></switch>
```

Here, the type of the switch is `toggle` instead of `circle`.

The last subsection is the temperature log graphic, which can be used to show to the user how the water temperature has been changed during the last 20 cycles. The code to implement this part is as follows:

```
<display id="temp_graph" type="graph" scale="range"
autoscale="true" mode="ch1" power_onload="true">
```

```
        <channel href="log_temp.php" refresh="60" sweep="0.005"
    frequency="20"></channel>
        </display>
```

Note that in this case, we need a special handler to generate the graphical points representing the water temperature. This handler is called `log_temp.php` and it's specified in the `href` parameter of the `channel` entry, while the other parameters define the refresh time in seconds (`refresh="60"`) and the scaling of the graph (`sweep` and `frequency`). See the `Drinks` documentation pages for further information about these parameters.

At each refreshing time, the `log_temp.php` script is called, and it will return the points sequence to be displayed to the display widget. To discover how it happens, we have to move to the next section. But before doing it, let me show you how the web control panel we just presented is displayed on my PC:

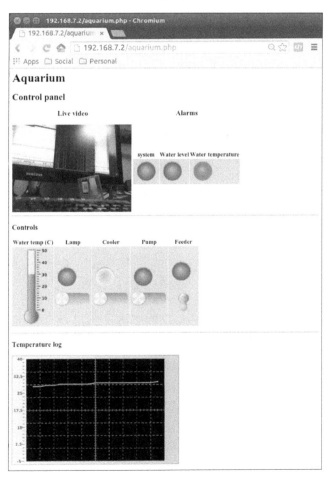

Handling the control panel

In the previous section, we discovered that the web control panel needs some handlers to send/receive data from the underlying system. In particular, we said that the input/output managers talk with the handler.php script, while the temperature log needs the log_temp.php script to get graph data. Let's see how these scripts are written.

The code in the handler.php script is as follows:

```php
<?php

require("db.php");

# Open the DB
db_open("aquarium_mon");

if (count($_GET) > 0) {
   # Input section
        db_set_status("force_cooler", $_GET["force_cooler"]);
        db_set_status("force_pump", $_GET["force_pump"]);
        db_set_status("force_lamp", $_GET["force_lamp"]);

    if ($_GET["force_feeder"])
        db_set_status("force_feeder", 1);
}

# Output section
$values["alarm_sys"] = db_get_status("alarm_sys");
$values["alarm_level"] = db_get_status("alarm_level");
$values["alarm_temp"] = db_get_status("alarm_temp");

$values["water"] = db_get_status("water");
$values["cooler"] = db_get_status("cooler");
$values["pump"] = db_get_status("pump");
$values["lamp"] = db_get_status("lamp");
$values["feeder"] = db_get_status("force_feeder");

$values["force_feeder"] = 0;

echo json_encode($values);
?>
```

The script can be found in the `chapter_03/handler.php` file in the book's example code repository.

This script has an input section to manage the input widgets and an output section to manage the output ones. In the input section, we simply get the status of each input widget and then store it in the `status` table. The only exception is for the `force_feeder` variable, which is recorded only if its status is 1 due to the the fact that it will be cleared when the feeder is actually enabled in the next cycle (again, the meaning of *next cycle* will be explained soon).

In the output section, we simply get the status of each status variable from the database and then store it into an array that will be returned to the `Drinks` toolkit by using the `json_encode()` function. The only note to be highlighted here is the fact that once the `force_feeder` switch is moved to the high state, its status is recorded, and then is cleared again, just to simulate the fact that this is not a normal on/off switch but just an *impulse-on* switch.

On the other hand, as said just now, the `log_temp.php` script just has to return a list of points. The following is the code:

```php
<?php

require("db.php");

# Open the DB
db_open("aquarium_mon");

# Get the last 20 points
$query = "SELECT v FROM temp_log ORDER BY t DESC LIMIT 20";
$ret = mysql_query($query);
if (!$ret)
  die();

$data = array();
$n = 0;
while ($row = mysql_fetch_array($ret)) {
  array_unshift($data, $row["v"]);
  $n++;
}

if ($n < 20)
  echo json_encode(array_merge(array_fill(0, 20 - $n, 0), $data));
else
  echo json_encode($data);
?>
```

The script can be found in the `chapter_03/log_temp.php` file in the book's example code repository.

The script simply selects the last 20 recorded points from the `temp_log` table and then stores them into the `data[]` array, adding some zeros at the beginning just in case there exists less then 20 stored temperature values. The `array_unshift()` function is used to put each new extracted value at the beginning of the array due to the fact that the SELECT statement returns the data in a reversed order.

Now, the last step is to put all these scripts together into the web browser root directory. The `/var/www` directory on my BeagleBone Black looks like:

```
root@beaglebone:~/chapter_03# ls /var/www/
Display.js   Knob.js   Slider.js   aquarium.css   log_temp.php
Drinks.js    Led.js    Switch.js   aquarium.php   handler.php
```

 The CSS can be found in the `chapter_03/aquarium.css` file in the book's example code repository. The code is not presented here since it's very trivial and not strictly needed for the understanding of the project.

Knowing the internal state-machine

Now that the control panel has been correctly set up, we have to take a look at the internal *state-machine*, that is, the procedure that at each cycle collects all the environment data and then decides what to do according to its internal state and the new environment status.

Our machine is implemented in the `chapter_03/aquarium_mon.php` file in the book's example code repository. The following are several code snippets of its `daemon_body()` function (the real core of the program.)

At its beginning, the function looks like:

```
function daemon_body()
{
    global $loop_time;
    global $sensors;

    $pump_time = strtotime("now");
    $feeder_time = strtotime("now");

    # The main loop
```

```
dbg("start main loop (loop_time=${loop_time}s)");
while (sleep($loop_time) == 0) {
dbg("loop start");

    $alarm_sys = 0;
```

At its beginning, the function initializes some variables and then starts its main loop where the first step is to get the water temperature, since, according to this value, many jobs need to be carried out!

Note also that the while() statement executes each time the sleep($loop_time) function, that is, each loop_time seconds starts a new *machine cycle*, and all the variables are modified according to the environment data read and the user inputs.

Then, the code continues reading the temperature as follows:

```
#
# Temperature management
#

$ret = temp_get();
if ($ret === false) {
    err("unable to get temperature!");
    $alarm_sys = 1;
}
$temp = $ret;
dbg("t=$temp");

# Save status
db_set_status("water", $temp);

#
# Check alarms
#

$water_temp_min = db_get_config("water_temp_min_alarm");
$water_temp_max = db_get_config("water_temp_max_alarm");
$val = ($temp < $water_temp_min ||
   $temp > $water_temp_max) ? 1 : 0;
db_set_status("alarm_temp", $val);

# Store the result into the proper log table
db_log_var("temp_log", $temp);

$water_level = get_water_level();
db_set_status("alarm_level", $water_level);
```

The `temp_get()` function reads the water temperature by reading the corresponding `w1_slave` file. It stores this value in the `temp` variable, and then some alarms are checked against this new value. Note also that in this case, the `alarm_sys` variable can be set to `1` to signal whether an I/O error has occurred or not.

The `get_water_level()` function is used to read the GPIO connected with the water sensor, and its body looks as follows:

```
function get_water_level()
{
    global $gpios;

    return gpio_get($gpios["water"]) == 0 ? 1 : 0;
}
```

Note that, as the preceding code shows, the water sensor has an inversed logic.

Now, it's the lamp's turn:

```
#
# Lamp management
#

# The lamp is directly managed by the force_lamp switch

$lamp = db_get_status("force_lamp");

# Set the new status
set_lamp($lamp);
db_set_status("lamp", $lamp);
dbg("lamp %sactivated", $lamp ? "" : "de");
```

In the preceding snippet, we see that the lamp is turned on and off according to the user input, without any automatic mechanism from the system.

This is not true for the cooler. Its management code looks as follows:

```
#
# Cooler management
#

# The cooler must be enabled if temp > water_temp_max in order
# to try to reduce the temperature of the water...
$water_temp_max = db_get_config("water_temp_max");
$cooler = $temp > $water_temp_max ? 1 : 0;
```

```
# We must force on?
$force_cooler = db_get_status("force_cooler");
$cooler = $force_cooler ? 1 : $cooler;

# Set the new status
set_cooler($cooler);
db_set_status("cooler", $cooler);
dbg("cooler %sactivated", $cooler ? "" : "de");
```

The cooler status is set according to the `temp` value and the `water_temp_max` setting and, as last case, it can be forced on by the user if the `force_cooler` variable is set to `1`.

A similar functioning applies for the pump:

```
#
# Pump management
#

# The pump must be on for pump_t_on delay time and off for
    # pump_t_off delay time (if not forced of course...)
    $force_pump = db_get_status("force_pump");
    $pump = db_get_status("pump");
    $pump_interval = $pump ? db_get_config("pump_t_on") :
        db_get_config("pump_t_off");
    if ($force_pump ||
        strtotime("-$pump_time seconds") > $pump_interval) {
            $pump_time = strtotime("now");

            $pump = $force_pump ? 1 : !$pump;
    }

    # Set the new status
    set_pump($pump);
    db_set_status("pump", $pump);
    dbg("pump %sactivated", $pump ? "" : "de");
```

This time, the state on or off is set by a timeout and, again, the device can be forced on by the user input, that is, if the `force_pump` variable is set to `1`.

All the preceding three snippets call a proper function to set on or off the relative GPIO; for instance, the last one calls the set_pump() function to set the pump status. The function's body is as follows:

```
function set_pump($status)
{
    global $gpios;

    gpio_set($gpios["pump"], $status ? 0 : 1);
}
```

The other two functions are similar.

The last notice is for the feeder. This time the code is as follows:

```
#
# Feeder management
#

$force_feeder = db_get_status("force_feeder");
$feeder_interval = db_get_config("feeder_interval");
if ($force_feeder || (strtotime("-$feeder_time seconds") > $feeder_
interval)) {
    $feeder_time = strtotime("now");

    do_feeder();
    db_set_status("force_feeder", 0);
    dbg("feeder activated");
}
```

The feeder can be activated according to a timeout or by a user input; but, instead of the preceding example, the code calls the do_feeder() function to call the feeder.sh script presented, as shown in the preceding code, and then it must clear the force_feeder status variable to signal to the user that the feeder has been activated. The body of the do_feeder() function is as follows:

```
function do_feeder()
{
    system("feeder.sh &");
}
```

 The character & in the system() function is needed in order to create a dedicated process to execute the feeder.sh script.

Now, it's time to execute the script. On my system, I used the following command line to execute it in debugging mode:

```
root@beaglebone:~# ./aquarium_mon.php -d -f -l
aquarium_mon.php[3882]: signals traps installed
aquarium_mon.php[3882]: start main loop (loop_time=15s)
aquarium_mon.php[3882]: loop start
aquarium_mon.php[3882]: t=28.5
aquarium_mon.php[3882]: lamp deactivated
aquarium_mon.php[3882]: cooler activated
aquarium_mon.php[3882]: pump deactivated
aquarium_mon.php[3882]: feeder activated
aquarium_mon.php[3882]: loop end
...
```

 Note that on your system, the **PHP** support may not be installed. In this case, you can solve it by using the following command:
```
root@beaglebone:~# apt-get install php5 libapache2-mod-php5
```

Every 15 seconds, the script wakes up and executes all the preceding steps in a new *cycle* of the *state-machine*. Note that to work you must set up all the hardware as presented in this section.

Final test

To test the prototype, I turned on the board, and after the login, I set up the system by using the commands discussed before, or by using the chapter_03/SYSINIT.sh script in the book's example code repository, as follows:

```
root@beaglebone:~# ./SYSINIT.sh
done!
```

Then, I executed the aquarium_mon.php command as follows:

```
root@beaglebone:~# ./aquarium_mon.php -d -f -l
```

Also, I executed the video streamer with the following command:

```
root@beaglebone:~# LD_LIBRARY_PATH=/usr/local/lib/ mjpg_streamer -i
"input_uvc.so -y -f 10 -r QVGA" -o "output_http.so -w /var/www/"
```

Then, I pointed my browser to the `aquarium.php` file on the BeagleBone Black's IP address (that is, the URL `http://192.168.7.2/aquarium.php`) and the game is done!

Note that at this point, we can try to force some settings or try to change some configuration variables by using the `chapter_03/my_dump.sh` and `chapter_03/my_set.sh` scripts in the book's example code repository, as follows:

```
root@beaglebone:~# ./my_dump.sh config
n       v
feeder_interval      60
pump_t_off      60
pump_t_on       20
water_temp_max       27
water_temp_max_alarm      29
water_temp_min       20
water_temp_min_alarm      18
root@beaglebone:~# ./my_set.sh config water_temp_max_alarm 30.5
root@beaglebone:~# ./my_dump.sh config
n       v
feeder_interval      60
pump_t_off      60
pump_t_on       20
water_temp_max       27
water_temp_max_alarm      30.5
water_temp_min       20
water_temp_min_alarm      18
```

In the preceding setting, I changed the `water_temp_max_alarm` limit value just as an example, and you can do all the changes that you wish on your system.

Before ending this chapter, let me show you how this control panel looks on my smartphone:

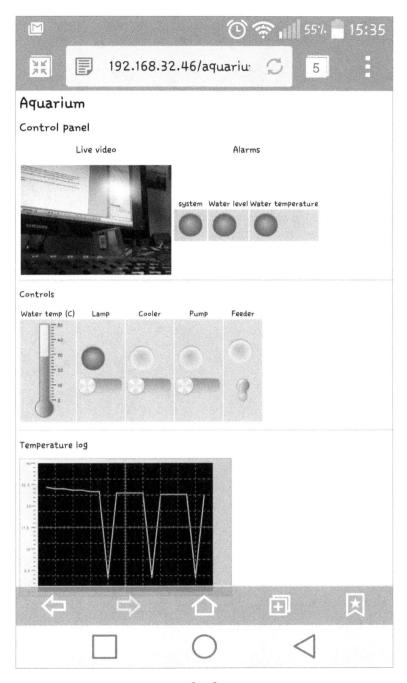

> The reader should notice that on the temperature log, there are three spikes due to the fact that during the temperature reading, the sensor returned an error. This issue can be fixed by repeating the reading two or three times before returning an error.

Summary

In this chapter, we discovered how to interface our BeagleBone Black to several devices with a different power supply voltage, and how to manage a 1-Wire device and a PWM one. Also, we presented the Drinks toolkit to realize a web control panel that can be used equally from a PC, smartphone, or tablet.

In the next chapter, we'll see how to realize a weather station that can store its collected data locally, which can not only show them in a nice manner on a web browser, but also can send its data to a Google Docs document!

Simply speaking, we're going to realize a simple **Internet-of-Things (IoT)** machine.

Google Docs Weather Station

4

In this chapter, we're going to take a look at a simple weather station that can also be used as an IoT device. This time, our BeagleBone Black will collect environment data and send them to a remote database in order to be reworked and presented into a shared environment.

Both local and remote data will be available in our preferred browser since, as it's a local system, we're going to use the `wfrog` tool, and as remote system, we're going to use a Google Docs spreadsheet.

The basics of functioning

In this project, our BeagleBone Black will collect the weather data through two sensors. But this time, instead of writing a dedicated software, we are going to use ready-made weather station software on our BeagleBone Black board to do the job. While on the remote side, we're going to use the well known Google Docs cloud system to store the data and then present them to the user.

In this manner, we can achieve a (*quasi*) professional result with minor effort!

In this scenario, our job is to connect the sensors, adapt the weather station software to our hardware in order to read the data from the sensors, and then add the proper code to send our data to a Google Docs spreadsheet.

Setting up the hardware

This time, the hardware setting is not very tricky since we just need two I^{2}C chips to get the basic environment data for our weather station, while all the complexity is in the software settings, since we need at least a 3.13 kernel to manage the sensors and a complete software toolchain to talk with the Google Docs system!

Maybe this is not the case with you, but my BeagleBone Black runs a kernel version 3.8 where some drivers are missing. That's why I decided to install a new distribution based on kernel release 3.13 on an external microSD so that I do not have to modify the default on-board eMMC settings.

In any case, just to set up the hardware, I can use the current running kernel where I can enable the I²C bus named I2C1 with the following command:

```
root@beaglebone:~# echo BB-I2C1 > /sys/devices/bone_capemgr.9/slots
```

If everything works well, you should see the following kernel activities on your board:

```
part_number 'BB-I2C1', version 'N/A'
slot #7: generic override
bone: Using override eeprom data at slot 7
slot #7: 'Override Board Name,00A0,Override Manuf,BB-I2C1'
slot #7: Requesting part number/version based 'BB-I2C1-00A0.dtbo'
slot #7: Requesting firmware 'BB-I2C1-00A0.dtbo' for board-name 'Override Board Name', version '00A0'
slot #7: dtbo 'BB-I2C1-00A0.dtbo' loaded; converting to live tree
slot #7: #2 overlays
omap_i2c 4802a000.i2c: bus 2 rev0.11 at 100 kHz
omap_i2c 4802a000.i2c: unable to select pin group
slot #7: Applied #2 overlays.
```

And a new device, /dev/i2c-2 should now be available:

```
root@beaglebone:~# ls -l /dev/i2c-2
crw-rw---T 1 root i2c 89, 2 Apr 23 20:23 /dev/i2c-2
```

Okay, now we can start adding the hardware to our BeagleBone Black and testing the connections with the current kernel.

 The reader can also take a look at the book *BeagleBone Essentials, Packt Publishing*, written by the author of this book, in order to have more information regarding how to manage BeagleBone Black's I²C buses needed to communicate with the sensors.

Setting up the temperature/humidity sensor

As a temperature and humidity sensor, I decided to use the device shown in the following image:

The device can be purchased at the following link (or by surfing the Internet): http://www.cosino.io/product/humidity-sensor.

The datasheet of this device is available at http://dlnmh9ip6v2uc.cloudfront.net/datasheets/BreakoutBoards/HTU21D.pdf.

This device is very simple. The I²C connections are as follows:

Pin	Temperature/humidity sensor pin
P9.4 - Vcc	+
P8.17 - CLK	CL
P8.18 - SDA	DA
P9.2 - GND	-

For further reading on the working of I²C bus, the reader can start with the Wikipedia article at http://en.wikipedia.org/wiki/I%C2%B2C.

Now, to verify the connections, we can use the `i2cdetect` command, as follows:

```
root@arm:~# i2cdetect -y -r 2
     0  1  2  3  4  5  6  7  8  9  a  b  c  d  e  f
00:          -- -- -- -- -- -- -- -- -- -- -- -- --
10: -- -- -- -- -- -- -- -- -- -- -- -- -- -- -- --
20: -- -- -- -- -- -- -- -- -- -- -- -- -- -- -- --
30: -- -- -- -- -- -- -- -- -- -- -- -- -- -- -- --
40: UU -- -- -- -- -- -- -- -- -- -- -- -- -- -- --
50: -- -- -- -- -- -- -- -- -- -- -- -- -- -- -- --
60: -- -- -- -- -- -- -- -- -- -- -- -- -- -- -- --
70: -- -- -- -- -- -- -- --
```

> Note that even if the I²C bus is named `I2C1` on the system, it must be addressed with the 2 ID number!

The string `UU` (or `40`) at the `0x40` address means that the device is connected! However, it may happen that you get no `UU` strings at all due to some hardware issues regarding this device. In this case, we can use the `i2cget` command as follows in order to force an I²C activity on the device:

```
root@beaglebone:~# i2cget -y 2 0x40 0xe7 0x02
```

Okay, the device is connected. But if you get the following output, you must recheck the connections:

```
root@beaglebone:~# i2cget -y 2 0x40 0xe7
Error: Read failed
```

> Note that you may need to disable the on-board pull-up resistors by clearing the soldered jumper on your sensor. In fact, the BeagleBone Black's I²C controller has the internal pull-up required by the I²C bus specifications, and under some circumstances, the pull-up on the sensor board may interfere with it.

Setting up the barometric sensor

As barometric sensor, I decided to use the device shown in the following image:

The device can be purchased at the following link (or by surfing the Internet): http://www.cosino.io/product/barometric_sensor.

The datasheet of this device is available at http://www.epcos.com/inf/57/ds/T5400.pdf, and a useful application note is at http://www.epcos.com/inf/57/ds/T5400.pdf.

This device has two interfaces: I²C and SPI. However, since the previous device was an I²C one, I decided to use the same interface. So, the connections must be done as reported in the following table, leaving the other pins unconnected:

Pin	Barometric sensor pin
P9.4 - Vcc	VCC
P8.17 - CLK	SCL/SCLK
P8.18 - SDA	SDA/MOSI
P9.2 - GND	GND

Note that we're going to connect both devices to the same I²C bus. For the moment, you can disconnect the previous sensor and then connect this one. But keep in mind that in the final configuration, all sensor devices are connected together with the same bus.

Now, to verify the connections, we can use the `i2cdetect` command, as follows:

```
root@arm:~# i2cdetect -y -r 2
     0  1  2  3  4  5  6  7  8  9  a  b  c  d  e  f
00:          -- -- -- -- -- -- -- -- -- -- -- -- --
10: -- -- -- -- -- -- -- -- -- -- -- -- -- -- -- --
20: -- -- -- -- -- -- -- -- -- -- -- -- -- -- -- --
30: -- -- -- -- -- -- -- -- -- -- -- -- -- -- -- --
40: -- -- -- -- -- -- -- -- -- -- -- -- -- -- -- --
50: -- -- -- -- -- -- -- -- -- -- -- -- -- -- -- --
60: -- -- -- -- -- -- -- -- -- -- -- -- -- -- -- --
70: -- -- -- -- -- -- -- 77
```

As the preceding command shows, the string `77` (or `UU`) at the `0x77` address means that the device is connected! This time, the device should be detected without any issue. So, if you do not get the preceding output, please consider rechecking your hardware connections.

The final picture

The following screenshot shows the prototype I realized to implement this project and to test the software. As you can see, the connections this time are very tricky.

Setting up the software

Now, it's time to play hard! We have to install a new kernel with specific patches in order to add the needed drivers. Then, we must set up our Google account in order enable the Google Docs API to manage a spreadsheet on the cloud. And, in the end, we must install and properly configure the weather station software we chose to collect the weather data.

Installing a new kernel

To install a new kernel, we must use a host PC where we use the following command to download the sources:

```
$ git clone git://github.com/RobertCNelson/bb-kernel.git
```

When finished, we must enter the `bb-kernel` directory and then check out the kernel, version 3.13:

```
$ git checkout am33x-v3.13
```

Now we should configure the compilation suite by generating a proper configuration file named `system.sh`, starting from the sample file as follows:

```
$ cp system.sh.sample system.sh
```

On my system, I modified the newly created `system.sh` file with the following settings:

```
CC=/usr/bin/arm-linux-gnueabihf-
MMC=/dev/sdd
```

The `MMC` variable is used by the installer tool (named `install_kernel.sh`), and it points to the device corresponding to the microSD where the BeagleBone system is installed.

 Warning! You must pay attention and be sure about the `MMC` define settings or the host machine may get damaged.

Now we must apply the patch into `chapter_04/0001-Add-support-for-I2C1-bus-and-the-connected-devices.patch` file in the book's example code repository in order to enable the I²C bus named `I2C1` and the drivers for the preceding sensors and to add the missing driver for the barometric sensor. The command is as follows:

```
$ git am --whitespace=nowarn  0001-Add-support-for-I2C1-bus-and-the-
connected-devices.patch
```

Note that the --whitespace=nowarn command line option is needed just in case your git system is configured to automatically fix up the whitespace errors, which is wrong in this case.

If everything works well, the following command should display the following text:

```
$ git log -1
commit 50949bd3a5c53d915dfdce8f790e3cfdd9ae702a
Author:     Rodolfo Giometti <giometti@hce-engineering.com>
AuthorDate: Wed Jun 24 21:58:50 2015 +0200
Commit:     Rodolfo Giometti <giometti@hce-engineering.com>
CommitDate: Wed Jun 24 22:06:06 2015 +0200

    Add support for I2C1 bus and the connected devices

    Signed-off-by: Rodolfo Giometti <giometti@hce-engineering.com>
```

Before starting the kernel compilation, let me spend a few words regarding this patch. It simply adds the following two patches:

```
$ ls patches/bbb-habp/
0001-iio-Add-t5403-barometric-pressure-sensor-driver.patch
0100-arm-am335x-bone-common.dtsi-enable-bus-I2C1-on-pins-.patch
```

The first patch is to add the driver for the barometric sensor and the second one is to enable the I²C bus labeled I2C1 and to define the connected devices. In particular, the second patch completes the steps shown in the following snippet:

```
diff --git a/arch/arm/boot/dts/am335x-bone-common.dtsi b/arch/arm/boot/
dts/am335x-bone-common.dtsi
index 5270d18..ba891ce 100644
--- a/arch/arm/boot/dts/am335x-bone-common.dtsi
+++ b/arch/arm/boot/dts/am335x-bone-common.dtsi
@@ -84,6 +84,13 @@
        >;
    };

+    i2c1_pins: pinmux_i2c1_pins {
+        pinctrl-single,pins = <
```

```
+              0x158 (PIN_INPUT_PULLUP | MUX_MODE2)      /* i2c1_sda.i2c1_sda */
+              0x15c (PIN_INPUT_PULLUP | MUX_MODE2)      /* i2c1_scl.i2c1_scl */
+         >;
+     };
+
      i2c2_pins: pinmux_i2c2_pins {
          pinctrl-single,pins = <
               0x178 0x73 /* (SLEWCTRL_SLOW | PIN_INPUT_PULLUP | MUX_MODE3)
uart1_ctsn.i2c2_sda */
@@ -295,6 +302,24 @@
     };
  };


+&i2c1 {
+    pinctrl-names = "default";
+    pinctrl-0 = <&i2c1_pins>;
+
+    status = "okay";
+    clock-frequency = <400000>;
+
+    htu21: htu21@40 {
+        compatible = "htu21";
+        reg = <0x40>;
+    };
+
+    t5403: t5403@77 {
+        compatible = "t5403";
+        reg = <0x77>;
+    };
+};
```

First, the preceding code defines the i2c1_pins group by selecting the proper pinmux settings, and then it enables the I2C1 bus, sets the correct bus frequency, and defines the proper drivers for the attached sensor devices.

Then, the patch adds the code needed to enable their application as follows:

```
$ git whatchanged -p -1 patch.sh
...
diff --git a/patch.sh b/patch.sh
index 83787f7..ed3a886 100644
--- a/patch.sh
+++ b/patch.sh
@@ -191,6 +191,12 @@ saucy () {
        ${git} "${DIR}/patches/saucy/0003-saucy-disable-stack-protector.
patch"
 }


+bbb_habp () {
+       echo "dir: bbb-habp"
+       ${git} "${DIR}/patches/bbb-habp/0001-iio-Add-t5403-barometric-
pressure-s
+       ${git} "${DIR}/patches/bbb-habp/0100-arm-am335x-bone-common.dtsi-
enable-
+}
+
 ###
 #arm
 deassert_hard_reset
@@ -211,4 +217,6 @@ boards

 saucy

+bbb_habp
+
 echo "patch.sh ran successful"
```

Also, as the last step, it enables the newly added driver into the default kernel's configuration:

```
$ git whatchanged -p -1 patches/defconfig
commit b9b954d37ed2722f7e85e9192d697bb79544ca78
Author:    Rodolfo Giometti <giometti@linux.it>
```

```
AuthorDate: Wed Jun 24 21:58:50 2015 +0200

Commit:     Rodolfo Giometti <giometti@linux.it>

CommitDate: Wed Jun 24 22:31:32 2015 +0200


    Add support for I2C1 bus and the connected devices


    Signed-off-by: Rodolfo Giometti <giometti@hce-engineering.com>


diff --git a/patches/defconfig b/patches/defconfig

index 7be0172..464301d 100644

--- a/patches/defconfig

+++ b/patches/defconfig

@@ -4529,6 +4529,7 @@ CONFIG_IIO_SYSFS_TRIGGER=m

 CONFIG_IIO_ST_PRESS=m

 CONFIG_IIO_ST_PRESS_I2C=m

 CONFIG_IIO_ST_PRESS_SPI=m

+CONFIG_T5403=m


 #

 # Temperature sensors
```

Okay, all modifications have been explained, and we can now start to compile the kernel with the following command:

```
$ ./build_kernel.sh
```

 This step and the subsequent ones are time consuming and require patience, so you should take a cup of your preferred tea or coffee and just wait.

After some time, the procedure will present the standard kernel configuration panel, and now we should verify that the needed drivers are enabled. You should navigate to the menu in **Device Drivers | Hardware Monitoring support** where the **Measurement Specialties HTU21D humidity/temperature sensors** entry should be selected as module (**<M>**), and in **Device Drivers | Industrial I/O support | Pressure sensors** where the **EPCOS T5403 digital barometric pressure sensor driver** entry should be selected as module too.

Then, exit from the menu and the kernel compilation will start. When it ends, the new kernel image is ready, and the following message should appear:

```
-----------------------------
Script Complete
eewiki.net: [user@localhost:~$ export kernel_version=3.13.10-bone12]
-----------------------------
```

> Note that when executing the build_kernel.sh file, you may get the following error message:
>
> ```
> $ ./build_kernel.sh
> + Detected build host [Ubuntu 14.04.3 LTS]
> + host: [x86_64]
> + git HEAD commit:
> [b00737d02a5b3567169a6c87311fec76a694fea6]
> Debian/Ubuntu/Mint: missing dependencies, please install:
> -----------------------------
> sudo apt-get update
> sudo apt-get install device-tree-compiler lzma lzop
> u-boot-tools libncurses5:i386 libstdc++6:i386
> -----------------------------
> * Failed dependency check
> ```
> In this case, you may resolve the problem by simply giving the preceding two suggested apt-get (or aptitude) commands.

Now, we can install it on the microSD using the installation tool:

```
$ ./tools/install_kernel.sh
```

Before updating the kernel, the tool asks whether the user is really sure about the device where the kernel must be placed. For example, on my system, I get the following output:

```
I see...
fdisk -l:
Disk /dev/sda: 500.1 GB, 500107862016 bytes
...
sdd        8:48   1   3.7G  0 disk
|-sdd1     8:49   1    12M  0 part   /media/giometti/BOOT
```

```
`-sdd2    8:50   1   3.7G  0 part  /media/giometti/rootfs
------------------------------
Are you 100% sure, on selecting [/dev/sdd] (y/n)?
```

My MMC variable is set to /dev/sdd; so, if I carefully take a look at the corresponding lines, I can verify that these are the right names of the BeagleBone Black's filesystem. So, I can safely answer yes by entering the y character.

> Note that the microSD should be a *class 10* and at least of 4 GB in size.

At the end of the command execution, we should get the output as follows:

```
This script has finished...
For verification, always test this media with your end device...
```

Now, just remove the microSD from the host machine and put it in your BeagleBone Black. Turn it on by keeping the user button pressed in order to force the boot from the microSD and, if everything works well, we can verify after the usual login that the new kernel is really running by using the following command:

```
# uname -a
Linux arm 3.13.10-bone9 #1 SMP Fri Nov 7 23:25:59 CET 2014 armv7l GNU/
Linux
```

> The reader can also take a look at the book *BeagleBone Essentials*, *Packt Publishing*, written by the author of this book, in order to have more information regarding how to install a newer kernel on an external microSD.

Okay, the new kernel is ready! Now we can verify that the needed drivers are also correctly loaded:

```
root@arm:~# lsmod | egrep '(t5403|htu21)'
t5403                3072  0
htu21                2385  0
industrialio        46516  3 t5403,ti_am335x_adc,kfifo_buf
```

The new devices can now be accessed through the *sysfs* interface. To get the current barometric pressure, we can use the following command:

```
root@arm:~# cat /sys/bus/iio/devices/iio\:device1/in_pressure_input
101.926000
```

The data is given in **kilopascal (kPa)**.

The temperature/humidity sensor can be accessed using the following commands:

```
root@arm:~# cat /sys/class/hwmon/hwmon0/device/humidity1_input
42988
root@arm:~# cat /sys/class/hwmon/hwmon0/device/temp1_input
27882
```

The humidity is given as a relative humidity percentage (m%RH), and the temperature is given in thousandths of Celsius degrees (m°C), so we have to divide both measurements by 1000 in order to have the relative humidity percentage (%RH) and the Celsius degrees (°C).

Running the weather station software

Now it's time to set up our weather station. To do so, as already stated, we decided to use an already made project instead of writing a new one. This is because there are tons of well-done weather station software existing that we can use to locally display the collected data in a better manner.

The software is the `wfrog` project.

> The home site of the project is at `https://code.google.com/p/wfrog/`.

To install it, we can get the sources from the online repository with the following command:

```
root@arm:~# svn checkout http://wfrog.googlecode.com/svn/trunk/ wfrog-read-only
```

> The `svn` command is located in the `subversion` package that can be installed by using the following command:
>
> `root@arm:~# aptitude install subversion`
>
> Note that a compressed archive of the program can be found in the `chapter_04/wfrog/wfrog-read-only.tgz` file in the book's example code repository.

After the download, we should go into the newly created directory `wfrog-read-only` and generate the `debian` package with the following commands:

```
root@arm:~# cd wfrog-read-only/
root@arm:~/wfrog-read-only# ./debian/rules binary
```

 Note that your system may miss some needed packages in order to be able to generate new `debian` packages. By using the following command, you should install whatever you need in order to do the job:

```
root@arm:~# aptitude install debhelper
```

If everything goes well, the `rules` command should display the following message:

```
dpkg-deb: building package `wfrog' in `../wfrog_0.8.2-1_all.deb'.
```

Then, to install the new packages, we can use the `gdebi` command, which will download all package's dependencies for us, as follows:

```
root@arm:~/wfrog-read-only# gdebi ../wfrog_0.8.2-1_all.deb
Reading package lists... Done
Building dependency tree
Reading state information... Done
Building data structures... Done
Building data structures... Done


Requires the installation of the following packages:
libxslt1.1 libyaml-0-2 python-cheetah python-lxml python-
pygooglechart python-serial python-support python-usb python-yaml
python2.6 python2.6-minimal
Web-based customizable weather station software
 wfrog is a software for logging weather station data and statistics,
 viewing them graphically on the web and sending them to a remote FTP
site.
 The layout and behaviour is fully customizable through an advanced
configuration system.
 It is written in python with an extensible architecture allowing new
station drivers to be written very easily.
```

```
wfrog supports many weather stations and is compliant with the WESTEP
protocol.
 Supported stations:
  * Ambient Weather WS1080
  * Davis VantagePro, VantagePro2
  * Elecsa AstroTouch 6975
  * Fine Offset Electronics WH1080, WH1081, WH1090, WH1091, WH2080, WH2081
  * Freetec PX1117
  * LaCrosse 2300 series
  * Oregon Scientific WMR100N, WMR200, WMRS200, WMR928X
  * PCE FWS20
  * Scientific Sales Pro Touch Screen Weather Station
  * Topcom National Geographic 265NE
  * Watson W8681
Do you want to install the software package? [y/N]:
...
Selecting previously unselected package wfrog.
(Reading database ... 40720 files and directories currently installed.)
Unpacking wfrog (from ../wfrog_0.8.2-1_all.deb) ...
Setting up wfrog (0.8.2-1) …
```

The gdebi command can be installed by using the following command:
`root@arm:~# aptitude install gdebi`

Okay, now the software is installed, but we still need to complete some steps before going further. The first one is to configure the system with a special simulator in order to verify that the web interface and the data collecting system are working correctly. To do so, we should execute the wfrog command as follows:

`root@arm:~# wfrog -S`

Note that, in case you get an error like the one shown as follows during the very first execution of the software, you may need the following patch to fix up the problem:

```
Traceback (most recent call last):
  File "/usr/bin/wfrog", line 132, in <module>
    settings = wflogger.setup.SetupClient().setup_
settings(SETTINGS_DEF, settings, settings_file)
  File "/usr/lib/wfrog/wflogger/setup.py", line 40, in
setup_settings
    if source == None:
UnboundLocalError: local variable 'source' referenced
before assignment
```

The patch is as follows:

```
root@arm:~/wfrog-read-only# svn diff
Index: wflogger/setup.py
===================================================================
--- wflogger/setup.py    (revision 973)
+++ wflogger/setup.py    (working copy)
@@ -35,6 +35,7 @@
        self.logger.debug('Current settings file:
'+str(source_file))
        self.logger.debug('New settings file:'+target_
file)
        defs = yaml.load( file(settings_def_file, 'r') )
+       source = None
        if source_file is not None:
            source = yaml.load( file(source_file, 'r') )
        if source == None:
```

Then, you simply need to rebuild the package.

The preceding patch can be found in the chapter_04/wfrog/0001-fix-setup.diff file in the book's example code repository.

Now, you should answer all the questions carefully selecting the 1) random-simulator option when the systems asks you to enter the driver for your station model, as follows:

```
Please enter the driver for your station model:
 1) random-simulator - Station Simulator
 2) vantagepro2 - Davis VantagePro
 3) wh1080 - Fine Offset WH1080 and compatibles
 4) wh3080 - Fine Offset WH3080 and compatibles
 5) wmr200 - Oregon Scientific WMR200
 6) wmr928nx - Oregon Scientific WMR928NX
 7) wmrs200 - Oregon Scientific WMRS200
 8) ws2300 - LaCrosse WS2300
 9) ws28xx - LaCrosse WS28xx
> 1
```

Once configured, you can start the weather station system by issuing the following two commands:

```
root@beaglebone:~# /etc/init.d/wflogger start
root@beaglebone:~# /etc/init.d/wfrender start
```

Then, the web interface can be accessed at `7680` port of the BeagleBone Black's IP address (usually `192.168.7.2`), as shown in the following screenshot:

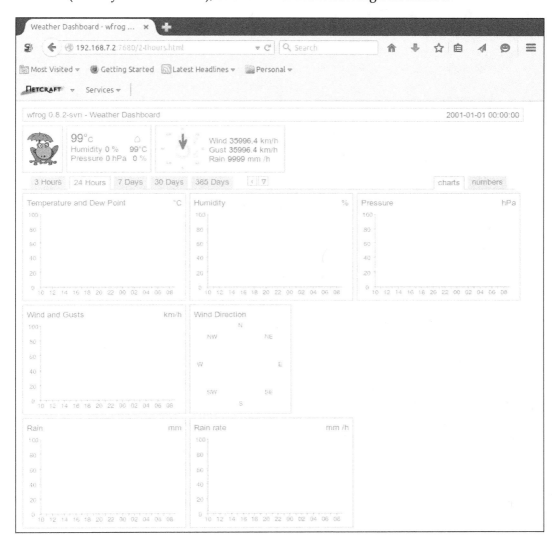

Okay, it works; but the system is now using a simulator, and we want it to use the data from the weather sensors we just installed! So, we have to add a new driver to our new weather station. To do so, we have to modify the sources just downloaded.

In the `wfrog-read-only/wfdriver/station/` directory, we have to add a new file called `bbb_habp.py`. The following shows a first snippet of its code with the lines that define a new class related to our new station:

```
import time
import logging
from wfcommon import units

class BBBhabpStation(object):

    '''

    Station driver for BeagleBone Black Home Automation Blueprints.

    [Properties]

    period [numeric] (optional):
        Polling interval in seconds. Defaults to 60.
    '''

    period=60

    logger = logging.getLogger('station.bbb_habp')

    name = 'BeagleBone Home Automation Blueprints weather station'
```

Then, the code defines the functions to read the environment data: the `get_press()` function reads the pressure, `get_temp()` reads the temperature, and `get_hum()` reads the humidity, as follows:

```
def get_press(self):
    f = open("/sys/bus/iio/devices/iio:device1/in_pressure_input", "r")
    v = f.read()
    f.close()

        return float(v) * 10.0

def get_temp(self):
    f = open("/sys/class/hwmon/hwmon0/device/temp1_input", "r")
    v = f.read()
    f.close()

        return int(v) / 1000.0

def get_hum(self):
    f = open("/sys/class/hwmon/hwmon0/device/humidity1_input", "r")
    v = f.read()
```

```
    f.close()

    return int(v) / 1000.0
```

After this, the code defines the core function that has the task to generate all weather events by calling the `generate_event()` function with a proper argument. The events just generated are stored into the `e` variable, and we have to only fill its fields and send the data to the weather station with the `send_event()` function, as shown in the following code snippet:

```
def run(self, generate_event, send_event, context={}):
    while True:
        try:
            e = generate_event('press')
            e.value = self.get_press()
            send_event(e)
            self.logger.debug("press=%fhPa" % e.value)

        except Exception, e:
            self.logger.error(e)

        try:
            e = generate_event('temp')
            e.sensor = 0
            e.value = self.get_temp()
            send_event(e)
            self.logger.debug("temp=%fC" % e.value)

        except Exception, e:
            self.logger.error(e)

        try:
            e = generate_event('hum')
            e.sensor = 0
            e.value = self.get_hum()
            send_event(e)
            self.logger.debug("hum=%f%%RH" % e.value)

        except Exception, e:
            self.logger.error(e)

        try:
            e = generate_event('temp')
            e.sensor = 1
```

```
            e.value = self.get_temp()
            send_event(e)
            self.logger.debug("temp=%fC" % e.value)

        except Exception, e:
            self.logger.error(e)

        try:
            e = generate_event('hum')
            e.sensor = 1
            e.value = self.get_hum()
            send_event(e)
            self.logger.debug("hum=%f%%RH" % e.value)

        except Exception, e:
            self.logger.error(e)
```

The last lines are used to schedule the next period:

```
# pause until next update time
next_update = self.period - (time.time() % self.period)
time.sleep(next_update)
```

 The preceding code is reported in the chapter_04/wfrog/ bbb_habp.py file in the book's example code repository.

Now, to finish the job, we have to patch the wfrog-read-only/wfdriver/ station/__init__.py file as follows:

```
root@arm:~/wfrog-read-only# svn diff wfdriver/station/__init__.py
Index: wfdriver/station/__init__.py
===================================================================
--- wfdriver/station/__init__.py    (revision 973)
+++ wfdriver/station/__init__.py    (working copy)
@@ -19,6 +19,7 @@
 import yaml


 import simulator
+import bbb_habp
 import wmrs200
 import wmr928nx
```

```
 import wmr200
@@ -66,6 +67,10 @@
     yaml_tag = u'!ws28xx'
 auto.stations.append(ws28xx)


+class YamlWS28xxStation(bbb_habp.BBBhabpStation, yaml.YAMLObject):
+    yaml_tag = u'!bbb_habp'
+auto.stations.append(bbb_habp)
+
 class YamlRandomSimulator(simulator.RandomSimulator, yaml.YAMLObject):
     yaml_tag = u'!random-simulator'
 auto.stations.append(simulator)
```

In this manner, we are saying to the wfrog system that a new station has been added.

 The patch is held in the chapter_04/wfrog/0002-add-bbb_habp-station.diff file in the book's example code repository.

After all the preceding modifications, we have to regenerate the package as just shown and reconfigure it by rerunning the configuration program and selecting the new driver as follows:

```
root@arm:~# wfrog -S
This is the setup of wfrog 0.8.2-svn user settings that will be written
in /etc/wfrog/settings.yaml


Please enter the driver for your station model:
 1) bbb_habp - BeagleBone Home Automation Blueprints weather station
 2) random-simulator - Station Simulator
 3) vantagepro2 - Davis VantagePro
 4) wh1080 - Fine Offset WH1080 and compatibles
 5) wh3080 - Fine Offset WH3080 and compatibles
 6) wmr200 - Oregon Scientific WMR200
 7) wmr928nx - Oregon Scientific WMR928NX
 8) wmrs200 - Oregon Scientific WMRS200
 9) ws2300 - LaCrosse WS2300
[random-simulator] > 1
```

Let me remind you that in order to regenerate the package, you have to execute the following two commands:

```
root@arm:~/wfrog-read-only# ./debian/rules binary
root@arm:~/wfrog-read-only# gdebi ../wfrog_0.8.2-1_
all.deb
```

Note that this time, a new entry named `bbb_habp` is added, so just select it and reconfigure the system as needed.

When all modifications are in place, we have to stop the running `wfrog` tasks:

```
root@arm:~# /etc/init.d/wflogger stop
[ ok ] Stopping wfrog logger - Weather Station Software : wfrog.
root@arm:~# /etc/init.d/wfrender stop
[ ok ] Stopping wfrog renderer - Weather Station Software : wfrender.
```

Then, we can safely clear the files where `wfrog` holds the weather data with the following command:

```
root@arm:~# rm /var/lib/wfrog/wfrog-current.xml /var/lib/wfrog/wfrog.csv
```

And then, we can restart `wfrog` tasks as follows:

```
[....] Starting wfrog logger - Weather Station Software : wfrogStarting /
usr/lib/wfrog/bin/wfrog...
Detaching to start /usr/lib/wfrog/bin/wfrog...done.
. ok
root@arm:~/chapter_04# /etc/init.d/wfrender start
[ ok ] Starting wfrog renderer - Weather Station Software : wfrender.
```

Now, if you take the files in the `/var/lib/wfrog` directory under control, you should see that they will be repopulated with the new data from the sensors.

Adding the Google Docs API

Our weather station is now fully functional, but as stated at the beginning of this chapter, we want more—we want a weather station able to save its data over the network on a cloud system. And to do so, we have decided to use a Google Docs spreadsheet.

The idea is to obtain a worksheet with the current weather data and the historical ones saved in it where all the data are updated periodically. So, let's see how to do this.

The API to be used for this task is implemented by the `gspread` tool, which can be installed into our BeagleBone Black by using the following commands:

```
root@arm:~# aptitude install python-pip python2.7-dev libffi-dev
root@arm:~# pip install --upgrade cffi cryptography PyOpenSSL
oauth2client gspread
```

 The home site of the project is at `https://github.com/burnash/gspread`.

After the installation of all the preceding packages, we need to prepare a Google account. In this project, I used my own account, so I simply get access to my Google Docs page. Then, I create a new spreadsheet named `bbb_weather`.

 Please refer to the Google Docs documentation at `https://support.google.com/docs` for further information about the Google Doc usage.

Once created, we have to publish the spreadsheet in such a way that we can get access to it from a remote computer. To do so, we must follow the instructions at `http://gspread.readthedocs.org/en/latest/oauth2.html`, where the OAuth2 authorization system is explained. The following is a little list of the needed steps directly from that page:

1. Navigate to Google Developers Console (`https://console.developers.google.com/project`) and create a new project (or select the one you have).

2. Under **API & auth** in the API, enable **Drive API**.

3. Navigate to **Credentials** and click on **Create new Client ID**.

4. Select **Service account**. Clicking on **Create Client ID** will generate a new public-private key pair. You will automatically download a JSON file with the following data inside:

```
{
    "private_key_id": "2cd ... ba4",
    "private_key": "-----BEGIN PRIVATE KEY-----\nNrDyLw ...
                        jINQh/9\n-----END PRIVATE KEY-----\n",
    "client_email": "473 ... hd@developer.gserviceaccount.com",
    "client_id": "473 ... hd.apps.googleusercontent.com",
    "type": "service_account"
}
```

5. Go to Google Sheets and share your spreadsheet with an e-mail you have in your `json_key['client_email']`. Otherwise, you'll get a `SpreadsheetNotFound` exception when trying to open it.

 In the next code examples, my Google credentials are stored in the `Project-9a372e9e20e6.json` file which, for security reasons, is not reported in the book's example code repository.

Now, to test if everything has been correctly set up, you can use the following command to create a void sheet in the newly-created spreadsheet:

```
root@arm:~# ./create_new.py
```

 You can find the preceding file in the `chapter_04/wfrog/create_new.py` file in the book's example code repository.

If you get no errors, you should find a new sheet named `BBB weather`, as shown in the following screenshot:

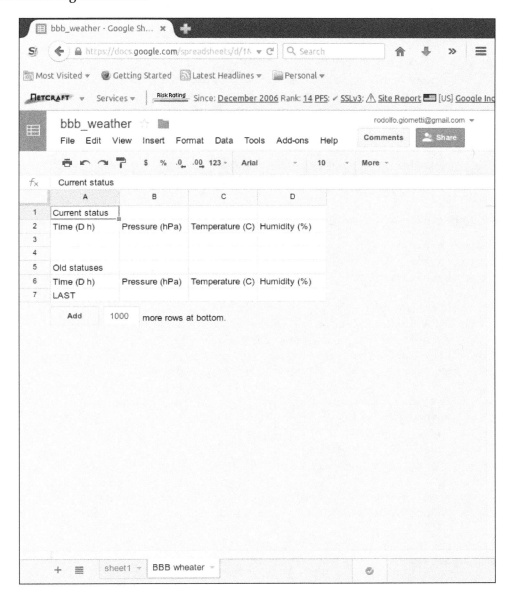

Note that when executing the preceding command, you may get the following error:

ImportError: No module named httplib2

In this case, you may resolve installing the missing `python-httplib2` package with the following command:

root@arm:~# aptitude install python-httplib2

Or you may get the following error:

oauth2client.client.CryptoUnavailableError: No crypto library available

In this case, the solution is to install the missing `python-crypto` package with the following command:

root@arm:~# aptitude install python-crypto

The code in the `create_new.py` file is quite simple, as follows:

```python
import gspread
import json
from oauth2client.client import SignedJwtAssertionCredentials

# Load the credentials
json_key = json.load(open('Project-9a372e9e20e6.json'))
scope = ['https://spreadsheets.google.com/feeds']
credentials = SignedJwtAssertionCredentials(json_key['client_email'],
json_key['private_key'], scope)

# Ask for authorization
gc = gspread.authorize(credentials)

# Open the "bbb_weather" spreadsheet
sh = gc.open("bbb_weather")

# Add a new worksheet named "BBB weather" with size of 7x4 cells
wks = sh.add_worksheet(title="BBB weather", rows="7", cols="4")

# Setup the "current status" part
wks.update_acell('A1', 'Current status')

wks.update_acell('A2', 'Time (D h)')
wks.update_acell('B2', 'Pressure (hPa)')
wks.update_acell('C2', 'Temperature (C)')
```

```
wks.update_acell('D2', 'Humidity (%)')

# Setup the "old statuses" part
wks.update_acell('A5', 'Old statuses')

wks.update_acell('A6', 'Time (D h)')
wks.update_acell('B6', 'Pressure (hPa)')
wks.update_acell('C6', 'Temperature (C)')
wks.update_acell('D6', 'Humidity (%)')

wks.update_acell('A7', 'LAST')
```

First of all, you should note that you must provide your own credentials to the `json.load()` function (that is, your `Project-xxxxxxxxxxxx.json` file obtained just now). Then, note that after opening the `bbb_weather` spreadsheet, we simply add a new worksheet called `sh.add_worksheet()` by using the `gc.open()` function. Then, we set up the cells content with the `wks.update_acell()` method.

Now, the reader may be curious regarding why the cell A7 holds the LAST string. Be patient, this will be explained soon!

Okay, now we need a way to send the collected data to our new worksheet. As seen before, the `wfrog` program stores its data into two files: `wfrog-current.xml`, which holds the current values, and `wfrog.csv`, which holds the historical data. To do the job, we can use the code stored in the `chapter_04/send_data.py` file in the book's example code repository. The following are some relevant snippets. The first part of this file is the same as the `create_new.py` command, so we can skip the beginning till the opening of the `bbb_weather` spreadsheet:

```
# Open the "bbb_weather" spreadsheet
sh = gc.open("bbb_weather")

# Select the worksheet named "BBB weather"
wks = sh.worksheet("BBB weather")
Then we can parse the XML file and extract the current status data to
be send over the network:
#
# Send data to Google Docs
#

# Parse the XML file holding the current weather status
xmldoc = minidom.parse('/var/lib/wfrog/wfrog-current.xml')

# Extract the data
```

```
time_obj = xmldoc.getElementsByTagName('time')
time = time_obj[0].firstChild.nodeValue
press_obj = xmldoc.getElementsByTagName('pressure')
press = float(press_obj[0].firstChild.nodeValue)
temp_obj = xmldoc.getElementsByTagName('temp')
temp = float(temp_obj[0].firstChild.nodeValue)
hum_obj = xmldoc.getElementsByTagName('humidity')
hum = float(hum_obj[0].firstChild.nodeValue)
print "current: %s press=%f temp=%f hum=%f" % (time, press, temp, hum)
```

Once extracted, this data can be sent on the relative cells:

```
# Update the current status
wks.update_acell('A3', time)
wks.update_acell('B3', press)
wks.update_acell('C3', temp)
wks.update_acell('D3', hum)
```

Now it's time to save the historical data. This is where the LAST string comes to help us! First of all, we have to parse the .csv file:

```
# Parse the CSV file holding the old weather statuses
csvfile = open('/var/lib/wfrog/wfrog.csv', 'rb')
reader = csv.reader(csvfile, delimiter=',')

# Skip the headers
headers = reader.next()
```

Then, we ask for the row number holding the LAST string inside our worksheet by using the wks.find("LAST").row method. Since the LAST string is in the row number 7 at the very beginning, we can find how many rows to skip in the wfrog.csv file in order to find the new data to be stored, as follows:

```
# Find the "LAST" string where to insert data to
last = wks.find("LAST").row - 7
print "last saved row was %d" % last

# Skip already read row
for i in range(0, last):
    dummy = reader.next()
```

Now we can extract the data and save them in the worksheet by using the `wks.insert_row(data, n)` function that stores them in the *nth* row:

```
# Start saving not yet saved data
for row in reader:
    time = row[1]
    press = float(row[11])
    temp = float(row[2])
    hum = float(row[3])
    print "old: %s press=%f temp=%f hum=%f" % (time, press, temp, hum)

    # Add a new line with an old status
    wks.insert_row([time, press, temp, hum], 7 + last)
    last += 1
```

To test the code, we can execute the `send_data.py` command as follows:

```
root@arm:~# ./send_data.py
current: 2015-06-27 13:56:00 press=1026.354367 temp=29.083000
hum=44.537000

last saved row was 0
old: 2015-06-27 12:24:32 press=1026.700000 temp=29.200000 hum=49.600000
old: 2015-06-27 12:35:00 press=1026.700000 temp=29.500000 hum=50.100000
old: 2015-06-27 12:45:00 press=1026.600000 temp=29.500000 hum=48.800000
old: 2015-06-27 12:55:00 press=1026.700000 temp=29.400000 hum=48.400000
old: 2015-06-27 13:05:00 press=1026.700000 temp=29.300000 hum=47.500000
old: 2015-06-27 13:15:00 press=1026.600000 temp=29.200000 hum=48.100000
old: 2015-06-27 13:25:00 press=1026.500000 temp=29.100000 hum=45.900000
old: 2015-06-27 13:35:00 press=1026.500000 temp=28.700000 hum=47.100000
```

Note that in the preceding file, you have to modify the following line according to your JSON filename!

```
json_key = json.load(open('Project-9a372e9e20e6.json'))
```

The program correctly detects that no historical data was saved before, and it starts to save new data from the very beginning. My worksheet now looks like what is shown in the following screenshot:

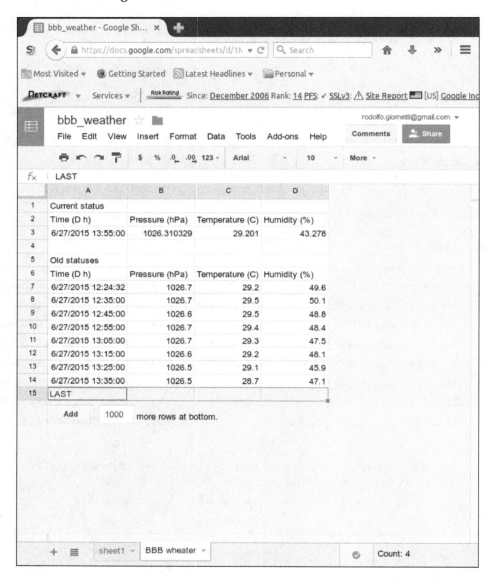

Now, the LAST string points to the row 15, so if we wait for new data and rerun the command, we get the following:

```
root@arm:~# ./send_data.py
```

```
current: 2015-06-27 13:51:00 press=1026.334273 temp=29.276000
hum=46.871000

last saved row was 8

old: 2015-06-27 13:46:00 press=1026.500000 temp=28.900000 hum=48.200000
```

As we can see in the following screenshot, our program has just saved the new data at the right position:

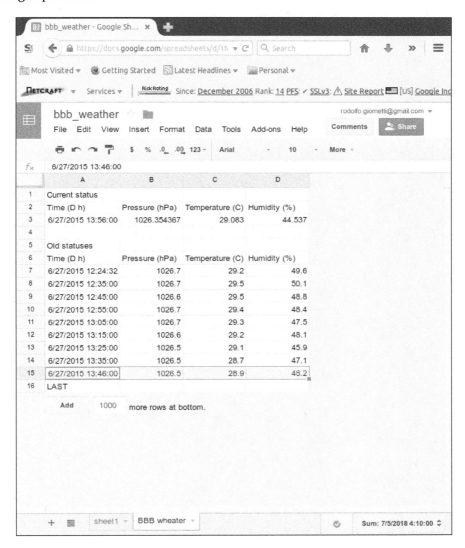

To automate these steps, we can use the `cron` daemon, scheduling the execution of the `send_data.py` program at the desired delays.

Final test

This time, the final test needs more time than other projects since we must collect several data to get suitable graphics. So, we execute the wfrog tasks as shown in the preceding screenshot. Then, we leave them running for two days or more. For my test, the results are shown in the following two screenshots:

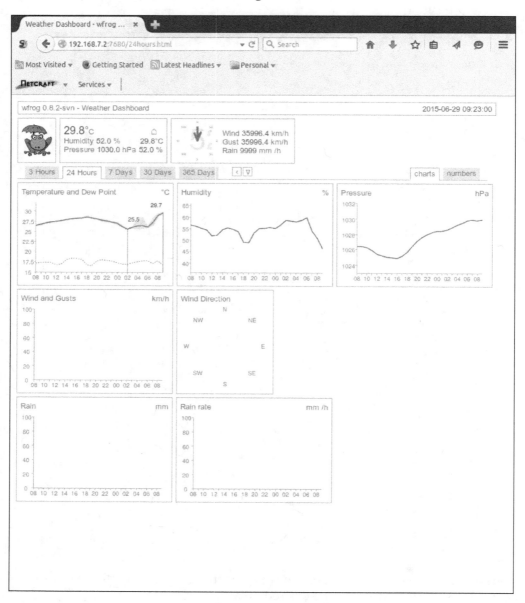

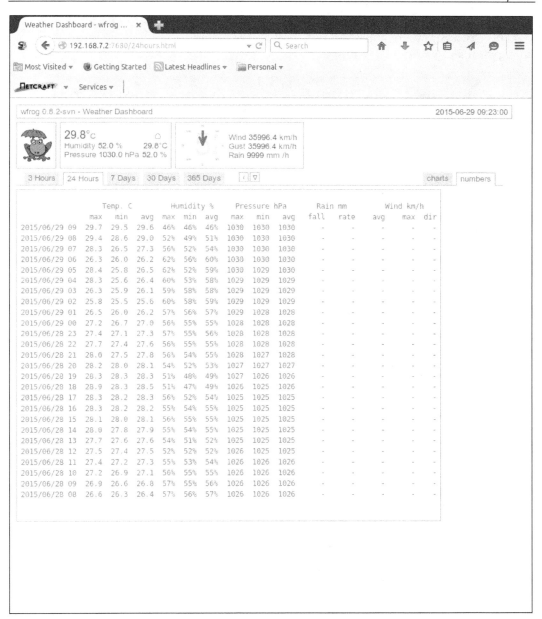

Then, we can verify that the preceding weather data has also been stored in the Google Docs spreadsheet by taking a look at our Google account. The following screenshot shows my test results:

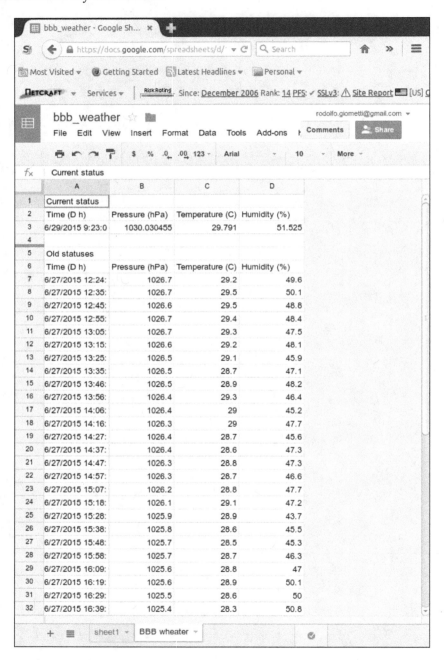

Summary

In this chapter, we discovered how to install a ready-to-use weather station software into our BeagleBone Black by adapting it to our hardware sensors and by installing a new kernel release with proper drivers. Then, we took a look at how to store data in a Google Docs spreadsheet for future processing.

In the next chapter, we'll continue to interact with a remote system in order to manage a laundry monitoring system. We're going to use the WhatsApp system to know when our washing machine has finished its duty.

5
WhatsApp Laundry Room Monitor

In this chapter, we'll see how to implement a laundry monitor room with several sensors capable of alerting the user directly on their WhatsApp account when a specific event occurs.

We'll see how to connect a sound sensor and a light sensor to our BeagleBone Black and then how we can monitor our washing machine with them. Also, we'll see how we can interact with the user directly on the user's smartphone by using a WhatsApp account in order to notify them of some events.

The basics of functioning

Let's assume that our laundry room is equipped with a washing machine and a lamp used by the user when they have to pick up the laundry. In these conditions, the BeagleBone Black can be equipped with some special sensors to detect when the washing machine has started or finished its job and when someone goes into the laundry room to pick up the washed clothes.

In this scenario, the BeagleBone Black should detect when the washing machine has been started by the user and then wait until the job has finished. At this point, the system can generate a WhatsApp message to alert the user that they have to pick up their clothes. When the user enters into the room, the light is turned on, and when they leave the room, the light is turned off. In this manner, our BeagleBone Black can detect when the user has done their job and then restart the cycle.

Setting up the hardware

As just stated, in this project we need two different kind of sensors: one to detect when the washing machine starts/stop, and one to detect when someone enters/exits the laundry room. The former task can be achieved by using a sound detector, that is, a device that is able to measure the environment sound level; while the latter task can be achieved by using a light sensor, that is, a device that is able to measure the environment light. Both these signals can be compared with thresholds in order to detect our relevant events.

When the washing machine is running, we should measure a high sound level for a long amount of time; while it is not running, the environment sound should be near to zero for a long time. On the other hand, we can assume that the person designed to pick up the washed clothes has to turn the light on in the laundry room, while the light is normally turned off when there is nobody in the room.

To help the user understand what happens inside the system, we can add two LEDs that can be turned on/off or put in a blinking mode with special meanings (in the next section, I'm going to explain these meanings in detail).

Setting up the sound detector

The device to detect the environment sound used in this project is shown in the following image:

The devices can be purchased at the following link (or by surfing the Internet): `http://www.cosino.io/product/sound-detector`.

The board is based on the amplifier LMV324 with the datasheet available at `http://dlnmh9ip6v2uc.cloudfront.net/datasheets/Sensors/Sound/LMV324.pdf`, while the board's schematic is available at `http://dlnmh9ip6v2uc.cloudfront.net/datasheets/Sensors/Sound/sound-detector.pdf`.

This device is very simple since it presents three outputs: the one labeled as **AUDIO** can be used to directly get the audio captured, while the output labeled **ENVELOPE** can be used to easily read the amplitude of sound by simply reading the analog voltage. The last output labeled **GATE** is a binary indication of the presence of the sound by using a fixed threshold (even if you can change it by changing the on-board resistors).

For our prototype, we can use the **ENVELOPE** output since we can read an analog voltage. Not only this, it allows us to set our own software threshold too. So, let's see the connections in the following table:

Pin	Sound Sensor Pin
P9.4 - Vcc	VCC
P9.39 - AIN0	R @ENVELOPE
P9.3 - GND	GND

As already mentioned in *Chapter 2, Ultrasonic Parking Assistant*, the ADC's input must be limited to 1.8V and since the Vcc level is 3.3V, we can use the voltage divider proposed there in order to scale the output voltage by a factor of 2. Be sure, then, that the maximum input level is not greater than 1.8V. So, the reader should not directly connect the *P9.39* pin with the sound detector; they should use the resistors connected as in *Chapter 2, Ultrasonic Parking Assistant*, to protect the BeagleBone Black's **ADC**.

Now, to verify that all the connections are okay, we enable the BeagleBone Black's ADCs by using the following command:

```
root@beaglebone:~# echo cape-bone-iio > /sys/devices/bone_capemgr.9/slots
```

These settings can be done by using the `bin/load_firmware.sh` script in the book's example code repository as follows:
```
root@beaglebone:~# ./load_firmware.sh adc
```

Then, we can read the captured sound envelope with the following command:

```
root@beaglebone:~# cat /sys/devices/ocp.3/helper.12/AIN0
24
```

If we try to speak while we rerun the command, we should get a higher value, as follows:

```
root@beaglebone:~# cat /sys/devices/ocp.3/helper.12/AIN0
201
```

So, the higher the environment sound, the higher the returned value.

> Let me remind you again that, as stated in *Chapter 1, Dangerous Gas Sensors*, the ADC can also be read by using another file's still in the `sysfs` filesystem with the following command:
>
> ```
> root@beaglebone:~# cat /sys/bus/iio/devices/
> iio:device0/in_voltage0_raw
> ```

Setting up the light sensor

The light sensor is the device shown in the following image:

> The devices can be purchased at the following link (or by surfing the Internet): http://www.cosino.io/product/light-sensor.
>
> The user guide of this device is available at http://www.phidgets.com/docs/1143_User_Guide.

As for the sound detector, this device has an analog output that can be used to measure the environment luminosity. According to the user guide, the luminosity can be obtained by using the following formula:

$Luminosity(lux) = e^{m*sensor\_output+b}$

Here, *sensor_output* is the raw value from the sensor, and the *m* and *b* are well-defined constants used to get a rough approximation. However, since we are only interested in measuring the light presence and not its precise intensity, we can use our own values or, to make it as simple as possible, the *sensor_output* value directly.

In the user guide, we also read that even if the device needs a 5V Vcc to function, its output value will not exceed 2.5V. So, considering that our BeagleBone Black's ADC maximum input value is 1.8V, we can use the voltage divider as above to scale down the output value by 2, thus being sure that the 1.8V threshold is satisfied.

The connections are as shown in the following table:

Pin	Light sensor cable
P9.6 - Vcc	red
P9.40 - AIN1	R @white
P9.1 - GND	black

Now, as done in the previous section for the sound detector, we can test the device by using the following command:

```
root@beaglebone:~# cat /sys/devices/ocp.3/helper.12/AIN1
386
```

However, if I put the device under a light, I get:

```
root@beaglebone:~# cat /sys/devices/ocp.3/helper.12/AIN1
528
```

Whereas if I cover the sensor with a cup of coffee, I get:

```
root@beaglebone:~# cat /sys/devices/ocp.3/helper.12/AIN1
79
```

So, as for the sound detector, the same rule exists: the higher the environment light intensity, the higher the returned value from the sensor.

Connecting the LEDs

To connect the LEDs, we can use the same circuitry that was used in *Chapter 1, Dangerous Gas Sensors*. The connections are reported in the following table:

Pin	LED color
P8.9 - GPIO69	R @red
P8.10 - GPIO68	R @yellow

To test the connections, we can use the following commands to set up the lines as outputs and then to set them into a high state:

```
root@beaglebone:~# ../bin/gpio_set.sh 68 out
root@beaglebone:~# ../bin/gpio_set.sh 69 out
root@beaglebone:~# echo 1 > /sys/class/gpio/gpio68/value
root@beaglebone:~# echo 1 > /sys/class/gpio/gpio69/value
```

If everything works well, you should see both LEDs turned on.

The final picture

The following image shows the prototype I realized to implement this project and to test the software:

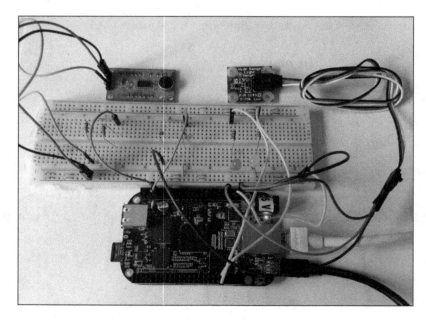

Nothing special to say here, apart from the fact that you must provide a network connection for your BeagleBone Black; otherwise, the WhatsApp alerting service will not work! As you can see, I used a normal Ethernet cable, but let me remind that you can also use a USB connection with the host, as mentioned in the *Preface*.

Setting up the software

This time, to implement the software of this prototype, we can use a state-machine with the following states and their relative transactions:

State	Description	Actions	Transaction conditions
IDLE	Idle state; the washing machine is not working.	• LED yellow off • LED red off	• If sound is detected, set t0=t and move state to SOUND.
SOUND	Sound detected! Keep monitoring the environment for a while.	• LED yellow is blinking • LED red is off	• If sound is detected and t-t0 > timeout, move to RUNNING.
RUNNING	Continuous sound detected so the washing machine has started its job.	• LED yellow is on • LED red is off • Alert the user	• If no sound is detected, set t0=t and move to NO_SOUND.
NO_SOUND	No more sound detected! Keep monitoring the environment for a while.	• LED yellow is on • LED red is blinking	• If no sound is detected and t-t0 > timeout, move to DONE. • If sound is detected, move to RUNNING.
DONE	No more sound for a long delay; the washing machine has finished its job.	• LED yellow is on • LED red is on • Alert the user	• If light is detected, set t0=t and move state to LIGHT.
LIGHT	Light detected! Keep monitoring the environment for a while.	• LED yellow is blinking • LED red is on	• If light is detected and t-t0 > timeout, move to ROOM. • If no light is detected, move to DONE.

State	Description	Actions	Transaction conditions
ROOM	Light is continuously on; someone has entered into the laundry room.	• LED yellow is off • LED red is on	• If no light is detected, set t0=t and move state to NO_LIGHT.
NO_LIGHT	No more light detected! Keep monitoring the environment for a while.	• LED yellow is off • LED red is blinking	• If no light is detected and t-t0 > timeout, move to IDLE.

The starting state is IDLE and the variable t holds the current time. t0 is used to address a starting time, while the timeout value can be fixed to a suitable amount of time in order to avoid false positive (so, you should try different values to suite your needs).

For each state, if any of the transaction conditions are not met, the state-machine assumes that no transaction must have been done at all and it remains in the original state.

Another representation of the preceding table is given by the following screenshot, where all the states of our machine are represented by circles and the state transactions are represented by arrows with a corresponding label holding the state transaction condition (the squares are just actions to be done before moving from one state to another). This representation more clearly shows the conditions we need to move from one state to another and how the states are connected to each other.

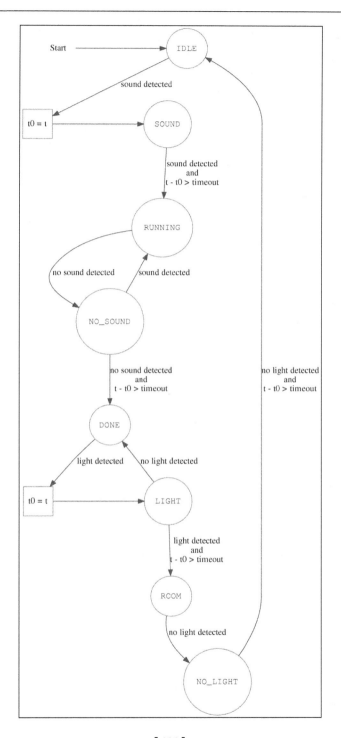

The sound detector manager

Okay, now we should try to understand how we can detect when the washing machine is running and when it's stopped. As already mentioned, the sound detector can help us distinguish these two states. In fact, by using the script in the `chapter_05/sample.sh` file in the book's example code repository, we can plot a graph of some samples taken from an ADC input. The script is simple and a snippet of the relevant code is as follows:

```
# Install the signals traps
trap sig_handler SIGTERM SIGINT

# Start sampling the data till a signal occours
echo "$NAME: collecting data into file sample.log..."

do_exit=false
t0=$(date '+%s.%N')
( while ! $do_exit ; do
        t=$(date '+%s.%N')
        v=$(cat $AIN_PATH/$dev)

        echo "$(bc -l <<< "$t - $t0") $v"

        # Sleep till the next period
        sleep $(bc -l <<< ".5 - $(date '+%s.%N') + $t")
done ) | tee sample.log

# Plot the data
echo "$NAME: done. Now generate the plot..."

gnuplot <<EOF
set terminal png size 800,600 enhanced font "Helvetica,20"
set output 'sample.png'
set autoscale
set nokey
set grid lw 1
show grid
set xlabel "\nTime"
set ylabel 'sample'
set xtics rotate
plot "sample.log" using 1:2 with lines
EOF

echo "$NAME: done. Data plotted into file sample.png"
```

The first part of the script is simply a `while` loop used to read the ADC data at more or less 500 ms (the script is in Bash, so don't expect too much precision from it). When the user strikes the *CTRL + C* keys, they generate a signal which is trapped by the `sig_handler` signal handler that simply sets the `do_exit` variable to `true`, as follows:

```
function sig_handler () {
        do_exit=true
}
```

The `tee` command is used to display the sampled data to the terminal and to save them in the `sample.log` file at the same time. Once the data are collected, we use the `gnuplot` tool to generate the graph in a similar way as done in *Chapter 1, Dangerous Gas Sensors*.

The following is a sample demo I did on my prototype. In the middle of the test, I discussed the letter *A* in order to produce a detectable sound level:

```
root@beaglebone:~# ./sample.sh AIN0
sample.sh: collecting data into file sample.log (press CTRL+C to stop)...
.046822125 29
.607965667 36
1.168452792 27
1.728863042 37
2.290465209 31
2.851453792 22
3.417320167 25
3.980918459 26
4.541227334 324
5.101803001 439
5.662116709 452
6.223465293 466
6.783610585 631
7.346517043 670
7.910204543 600
8.471078668 569
9.032048168 677
9.592383627 728
10.153342335 708
```

```
10.714916752 736
11.275682627 769
11.836266085 672
12.396825252 308
12.958963794 267
13.520244377 20
14.081049085 19
14.641610585 20
15.202588419 20
15.762929794 19
16.324602752 19
16.885479836 25
17.450904252 19
^Csample.sh: done. Now generate the plot...

    Rectangular grid drawn at x y tics
    Major grid drawn with linetype 0 linewidth 1.000
    Minor grid drawn with linetype 0 linewidth 1.000
    Grid drawn at default layer

sample.sh: done. Data plotted into file sample.png
```

As you can see in the preceding output where I just used my voice, I can easily distinguish sound absence or presence. However, the following screenshot, taken from the sample.png file generated by the preceding script, is more explicative:

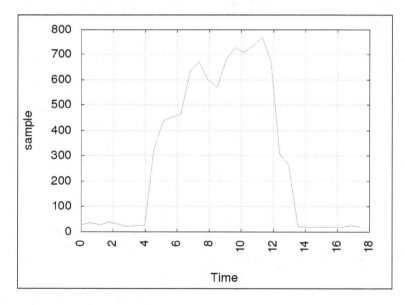

It's clear that just by using a threshold of 200, we can do the trick.

The light sensor manager

The light sensor functioning is very similar to the sound one, so we can use the same `sample.sh` script to get some samples from it. This time, I simulate the light absence/presence by simply covering the light sensor with a small cup.

The command used is as follows:

```
root@beaglebone:~# ./sample.sh AIN1
sample.sh: collecting data into file sample.log (press CTRL+C to stop)...
.046757875 78
.609375334 78
1.169878875 78
1.730529209 78
2.291640417 78
2.852100834 78
3.412606126 78
3.973172542 79
4.534430834 77
5.094665667 78
5.655394084 463
6.216107209 477
6.777377459 484
7.337617209 486
7.898274043 486
8.458853793 487
9.023789835 486
9.590632751 486
10.154009085 486
10.715376668 479
11.275998293 479
11.836406710 476
12.397433502 403
12.958216252 92
13.519537710 79
14.080658377 79
14.641473210 79
15.202038044 78
```

```
15.762509877 78
16.323857252 79
16.884874669 78
17.445492127 77
18.006021794 78
^Csample.sh: done. Now generate the plot...

    Rectangular grid drawn at x y tics
    Major grid drawn with linetype 0 linewidth 1.000
    Minor grid drawn with linetype 0 linewidth 1.000
    Grid drawn at default layer


sample.sh: done. Data plotted into file sample.png
```

And the corresponding plot is shown in the following screenshot:

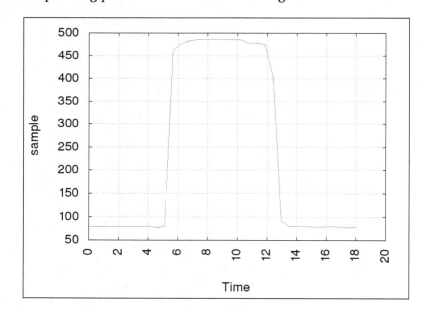

Even in this case, we can use a threshold of 200 to distinguish between the two states.

Controlling the LEDs

As already shown in *Chapter 1, Dangerous Gas Sensors*, or in *Chapter 2, Ultrasonic Parking Assistant*, there are two different manners of managing LED in a Linux-based system. The first is by using GPIO and the second is by using LED device; but, since our state-machine requires that the LEDs should blink, we should use the LED management method that allows us to use a trigger to get a blinking status.

Similarly, as done in *Chapter 2, Ultrasonic Parking Assistant*, we need a proper `.dts` file that the reader can find in the `chapter_05/BB-LEDS-C5-00A0.dts` file in the book's example code repository. After finding it, we have to compile it with the following command line:

```
root@beaglebone:~# dtc -O dtb -o /lib/firmware/BB-LEDS-C5-00A0.dtbo -b 0
-@ BB-LEDS-C5-00A0.dts
```

Now, we can enable it by using the following command:

```
root@beaglebone:~# echo BB-LEDS-C5 > /sys/devices/bone_capemgr.9/slots
```

And then, two new LEDs are now available in the system, as follows:

```
root@beaglebone:~# ls -d /sys/class/leds/c5*
/sys/class/leds/c5:yellow  /sys/class/leds/c5:red
```

Setting up the WhatsApp API

Now it's time to show you how we can interact with the **WhatsApp** service. In this project, we simply need to send messages to the user's account, but even this simple task needs us to accomplish several steps.

First of all, we must install some prerequisite packages into our BeagleBone Black, as follows:

```
root@beaglebone:~# aptitude install python python-dateutil python-argparse
```

Then, we have to install the package named `yowsup` that we can use to send our messages via WhatsApp:

```
root@beaglebone:~# pip install yowsup
```

> The wiki page of the `yowsup` tool is at `https://github.com/tgalal/yowsup/wiki`.

When the installation is finished, we can use the following command to get a sample configuration file:

```
root@beaglebone:~# yowsup-cli demos --help-config > yowsup-cli.config
```

The new file `yowsup-cli.config` should now hold the following lines:

```
root@beaglebone:~# cat yowsup-cli.config


############# Yowsup Configuration Sample ###########
#
# ====================
# The file contains info about your WhatsApp account. This is used during
# registration and login.
# You can define or override all fields in the command line args as well.
#
# Country code. See http://www.ipipi.com/help/telephone-country-codes.htm.
# This is now required.
cc=49
#
# Your full phone number including the country code you defined in 'cc',
# without preceding '+' or '00'
phone=491234567890
#
# You obtain this password when you register using Yowsup.
password=NDkxNTIyNTI1NjAyMkBzLndoYXRzYXBwLm5ldA==
####################################################
```

Lines starting with the character # are comments and they can be removed so the important lines are:

```
cc=39
phone=39XXXXXXXXX
id=
password=
```

 Note that the id= line may not be present.

In the preceding example, for privacy reasons I replaced my phone number with the x characters, but you have to put your phone number here in order to get access to the system.

> Note that you cannot use a phone number when you actually already use WhatsApp or else you're going to get into conflict with the WhatsApp client you are using on your smartphone. That's why I used a phone number when no WhatsApp services are active.
>
> Simply speaking, you don't need the WhatsApp client running on the phone that is receiving the SMS!

Once you have added a phone number, you can put it into the preceding `yowsup-cli.config` configuration file, leaving the lines with the `id` and `password` variables unassigned. Then, the following command must be executed:

```
root@beaglebone:~# yowsup-cli registration -r sms -c  yowsup-cli.config
```

After a while, the command should answer as follows:

```
INFO:yowsup.common.http.warequest:{"status":"sent","length":6,"method":"s
ms","retry_after":1805}
```

```
status: sent
retry_after: 1805
length: 6
method: sms
```

Then, you should receive an SMS on the phone with your number. You just need the information inside the message itself; in fact, the SMS should hold a message like `WhatsApp code 633-170`, so you have to use the following command to finish the registration:

```
root@beaglebone:~# yowsup-cli registration -R 633-170 -c yowsup-cli.
config
```

If everything works well, the preceding command should answer, as follows:

```
{"status":"ok","login":"39XXXXXXXXX","pw":"Kwf07sjuSz2J0Qwm3sBEtVNeBIk="
,"type":"new","expiration":1467142355,"kind":"free","price":"\u20ac0,89",
"cost":"0.89","currency":"EUR","price_expiration":1438319298}
```

```
status: ok
kind: free
```

```
pw: Kwf07sjuSz2J0Qwm3sBEtVNeBIk=
price: € 0,89
price_expiration: 1438319298
currency: EUR
cost: 0.89
expiration: 1467142355
login: 393292571400
type: new
```

The important information here is the password that we must use to correctly log in into our new WhatsApp account. The password is the field `pw`, so after putting this information into the `password` field of the configuration file, the new look of the `yowsup-cli.config` file should be as follows:

```
root@beaglebone:~# cat yowsup-cli.config
cc=39
phone=39XXXXXXXXX
id=
password=Kwf07sjuSz2J0Qwm3sBEtVNeBIk=
```

Now we are ready to log in into our new account and send messages from it! For example, the following command line can be used to send a message from the command line:

```
root@beaglebone:~# yowsup-cli demos -c yowsup-cli.config -s 39YYYYYYYYY
"Hello, it's your BeagleBone Black writing! :)"
WARNING:yowsup.stacks.yowstack:Implicit declaration of parallel layers in
a tuple is deprecated, pass a YowParallelLayer instead
INFO:yowsup.demos.sendclient.layer:Message sent
```

Yowsdown

Note that I used another phone number as the destination number, obscured as 39YYYYYYYYY, to distinguish it from the transmitter one used before.

> The warning message can be safely ignored.
>
> Also, it may happen that the first time you execute the command, you get no a "Message Sent" output. In this case, try to rerun the command.

Okay, now everything is in place and we just need to see how the state-machine can be implemented. So let's move to the next section.

The state-machine

Now that every subsystem has been set up, it's time to take a look at a possible implementation of the state-machine described before. The complete code is so simple that it has been developed in Bash and it can be found in the chapter_05/ state_machine.sh file in the book's example code repository. However, the following are some snippets of the relevant code.

The first snippet is about the configuration file reading, as follows:

```
SOUND_DEV="/sys/devices/ocp.3/helper.12/AIN0"
LIGHT_DEV="/sys/devices/ocp.3/helper.12/AIN1"

source ../lib/logging.sh
source ./config.sh

# Check the configuration settings. If not specified use default
values
[ -z "$TIMEOUT" ] && TIMEOUT=60
[ -z "$SOUND_TH" ] && SOUND_TH=500
[ -z "$LIGHT_TH" ] && LIGHT_TH=500
if [ -z "$WHATSAPP_USER" ] ; then
        err "you must define WHATSAPP_USER!"
        exit 1
fi
```

After some initial settings, the code sources the config.sh file that holds the system settings (see the last section for an example of this file), and then the settings variables are checked up. Then, the code continues defining the sensor's reading functions. In the following snippet, I reported only one of these functions since they are very similar:

```
function read_sound () {
        ret=0

        while [ -z "$v" ] ; do
                v=$(cat $SOUND_DEV)
        done
        [ "$v" -gt $SOUND_TH ] && ret=1

        echo -n $ret
}
```

The function simply reads the ADC and checks the datum against a specified threshold. The returned value is 0 or 1, according to the absence/presence of the sound or light. Note that in case of errors in reading the datum, the function retries the operation until a successful reading.

 Here, we should add a retries limit in order to avoid an infinite loop. But for the sake of simplicity, I decided to not implement it.

The LED's management section looks as follows:

```
function set_led () {
        name=$1
        val=$2

        case $val in
        on)
                echo none > /sys/class/leds/c5\:$name/trigger
                echo 255 > /sys/class/leds/c5\:$name/brightness
                ;;

        off)
                echo none > /sys/class/leds/c5\:$name/trigger
                echo 0 > /sys/class/leds/c5\:$name/brightness
                ;;

        blink)
                t=$((1000 / 2))

                echo timer > /sys/class/leds/c5\:$name/trigger
                echo $t > /sys/class/leds/c5\:$name/delay_on
                echo $t > /sys/class/leds/c5\:$name/delay_off
                ;;

        *)
                err "invalid LED status! Abort"
                exit 1
                ;;
        esac
}

function signal_status () {
        s=$1

        case $s in
        IDLE)
```

```
                set_led yellow off
                set_led red off
                ;;

        SOUND)
                set_led yellow blink
                set_led red off
                ;;

        RUNNING)
                set_led yellow on
                set_led red off
                ;;

        NO_SOUND)
                set_led yellow on
                set_led red blink
                ;;

        DONE)
                set_led yellow on
                set_led red on
                ;;

        LIGHT)
                set_led yellow blink
                set_led red on
                ;;

        ROOM)
                set_led yellow off
                set_led red on
                ;;

        NO_LIGHT)
                set_led yellow off
                set_led red blink
                ;;
        esac

        return
    }
```

The set_led function simply sets the LED status according to the system status passed by the signal_status function.

Note that the `signal_status` function can be implemented in a more compact manner (maybe by using an associative array), but this form is more readable.

Then, the code of the function to send the alerting messages via WhatsApp system is as follows:

```
function send_alert () {
        msg=$1

        dbg "user=$WHATSAPP_USER msg=\"$1\""
        yowsup-cli demos -c yowsup-cli.config -s $WHATSAPP_USER "$msg"

        return
}
```

Now, the core of the whole project is the `change_status` function. This is the function that implements the state-machine. It decides which is the new status according to the current one and the system's inputs:

```
function change_status () {
        status=$1
        sound=$2
        light=$3
        t0=$4

        t=$(date "+%s")

        dbg "status=$status sound=$sound light=$light t-t0=$(($t - $t0))"

        case $status in
        IDLE)
                if [ $sound -eq 1 ] ; then
                        echo SOUND
                        return
                fi
                ;;

        SOUND)
                if [ $sound -eq 1 -a $(($t - $t0)) -gt $TIMEOUT ] ; then
                        echo RUNNING
                        return
                fi
                if [ $sound -eq 0 ] ; then
                        echo IDLE
                        return
```

```
                fi
                ;;

RUNNING)
                if [ $sound -eq 0 ] ; then
                        echo NO_SOUND
                        return
                fi
                ;;

NO_SOUND)
                if [ $sound -eq 0 -a $(($t - $t0)) -gt $TIMEOUT ] ; then
                        echo DONE
                        return
                fi
                if [ $sound -eq 1 ] ; then
                        echo RUNNING
                        return
                fi
                ;;

DONE)
                if [ $light -eq 1 ] ; then
                        echo LIGHT
                        return
                fi
                ;;

LIGHT)
                if [ $light -eq 1 -a $(($t - $t0)) -gt $TIMEOUT ] ; then
                        echo ROOM
                        return
                fi
                if [ $light -eq 0 ] ; then
                        echo DONE
                        return
                fi
                ;;

ROOM)
                if [ $light -eq 0 ] ; then
                        echo NO_LIGHT
                        return
                fi
                ;;
```

```
        NO_LIGHT)
                if [ $light -eq 0 -a $(($t - $t0)) -gt $TIMEOUT ] ; then
                        echo IDLE
                        return
                fi
                if [ $light -eq 1 ] ; then
                        echo NO_LIGHT
                        return
                fi
                ;;

        *)
                err "invalid status! Abort"
                exit 1
                ;;
        esac

        # No status change!
        echo $status
}
```

You can verify that this function correctly implements the state-machine table (or graph) presented previously in this chapter.

At this point, the core of the main function looks like the following:

```
# Ok, do the job
dbg "using TIMEOUT=$TIMEOUT SOUND_TH=$SOUND_TH LIGHT_TH=$LIGHT_TH"

status="IDLE"
t0=0

signal_status $status
while sleep 1 ; do
        dbg "old-status=$status"

        # Read the sensors
        sound=$(read_sound)
        light=$(read_light)

        # Change status?
        new_status=$(change_status $status $sound $light $t0)
        if [ "$new_status" != "$status" ] ; then
                t0=$(date "+%s")
```

```
                    # Set the leds status
                    signal_status $new_status

                    # We have to send any alert?
                    case $new_status in
                    RUNNING)
                            # Send the message during SOUND->RUNNING
                            # transaction
                            # only
                            [ "$status" == SOUND ] && send_alert
        "washing machine is started!"
                            ;;

                    DONE)
                            # Send the message during NO_SOUND->DONE
                            # transaction
                            # only
                            [ "$status" == NO_SOUND ] && send_alert
        "washing machine has finished!"
                            ;;

                    *)
                            # Nop
                            ;;
                    esac
            fi
            dbg "new-status=$new_status"

            status=$new_status
    done
```

As you can see, the `main` function is just a big loop that periodically reads the sensor's inputs and then changes the system's internal status according to it, sending some alerts when needed and setting the LED's statuses accordingly.

Final test

To test the prototype, I used some tricks to simulate the washing machine and the light in the room. The washing machine can be easily simulated by an audio/video played on the host PC with a reasonable volume level, while the room light on/off status can be simulated by using a small cup to cover the light sensor.

To set up all peripherals and drivers, we can use all the preceding commands or the SYSINIT.sh script as follows:

```
root@beaglebone:~# ./SYSINIT.sh
done!
```

 This command can be found in the chapter_05/SYSINIT.sh file in the book's example code repository

As an initial state (IDLE), we should cover the light sensor (to simulate that the light is off) and we should stop the video/audio player (to simulate that the washing machine is off). Then, we have to set a low threshold level into the configuration file for both sound and light detection and a very short timeout (5 seconds) in order to speed up the test. The following is my configuration file:

```
root@beaglebone:~# cat config.sh
# Set the timeout value
TIMEOUT=5

# Set the sound threshold
SOUND_TH=200

# Set the light threshold
LIGHT_TH=200

# Set the Whatsapp account
WHATSAPP_USER=39YYYYYYYYYY
```

Then, I started the system, enabling all debugging messages on the terminal, by using the following command:

```
root@beaglebone:~# ./state_machine.sh -d -l
state_machine.sh: using TIMEOUT=5 SOUND_TH=200 LIGHT_TH=200
state_machine.sh: old-status=IDLE
state_machine.sh: status=IDLE sound=0 light=0 t-t0=1398295377
state_machine.sh: new-status=IDLE
state_machine.sh: old-status=IDLE
```

Note that the initial state is IDLE and nothing changes until no new events are detected.

In the next output listing, I'm going to use the . . . characters to skip non-relevant lines:

```
state_machine.sh: status=IDLE sound=0 light=0 t-t0=1398295379
...
state_machine.sh: status=IDLE sound=1 light=0 t-t0=1398295381
state_machine.sh: new-status=SOUND
state_machine.sh: old-status=SOUND
```

Now I'm going to simulate the following situation: first, I turn on the washing machine and wait for the end of its job. Then, I go to the laundry room to pick up my washed clothes. As already said before, I'm going to simulate the running washing machine with a video/audio player while the light on/off is simulated by uncovering/covering the light sensor with a cup.

Okay, the test begins. After a while, I start the video/audio player. So, a sound has been detected and the new state turns to SOUND:

```
state_machine.sh: status=SOUND sound=1 light=0 t-t0=1
...
state_machine.sh: status=SOUND sound=0 light=0 t-t0=4
state_machine.sh: new-status=IDLE
state_machine.sh: old-status=IDLE
```

Ouch! For a moment, the sound level went under the threshold, so we switched again to the IDLE state! This is correct because it may happen that the washing machine stops for a while. Here is where the timeout enters in action, that is, we have to select it for longer than all the possible washing machine's pauses:

```
state_machine.sh: status=IDLE sound=1 light=0 t-t0=1
...
state_machine.sh: old-status=SOUND
cat: /sys/devices/ocp.3/helper.12/AIN0: Resource temporarily unavailable
```

This an error during the reading of the ADC input, but the software is written to retry the faulty operation without problems:

```
state_machine.sh: status=SOUND sound=1 light=0 t-t0=4
...
```

```
state_machine.sh: status=SOUND sound=1 light=0 t-t0=6
```

```
state_machine.sh: user=393492432127 msg="washing machine is started!"
```

```
WARNING:yowsup.stacks.yowstack:Implicit declaration of parallel layers in
a tuple is deprecated, pass a YowParallelLayer instead
```

```
INFO:yowsup.demos.sendclient.layer:Message sent
```

```
Yowsdown
state_machine.sh: new-status=RUNNING
```

```
state_machine.sh: old-status=RUNNING
```

Good! When the timeout expires while we are into the SOUND state, it means that a continuous sound has been detected, so it means that the washing machine has started its job.

 Note that a more reliable implementation should use different timeouts to identify a specific transaction.

This is demonstrated in the following snippet:

```
state_machine.sh: status=RUNNING sound=1 light=0 t-t0=2
```

```
. . .
```

```
state_machine.sh: new-status=RUNNING
```

```
state_machine.sh: old-status=RUNNING
```

```
state_machine.sh: status=RUNNING sound=0 light=0 t-t0=8
```

```
state_machine.sh: new-status=NO_SOUND
```

```
state_machine.sh: old-status=NO_SOUND
```

Now, I have stopped the video/audio player and no sound has been detected, so we switch to the NO_SOUND state:

```
state_machine.sh: status=NO_SOUND sound=0 light=0 t-t0=1
```

```
. . .
```

```
state_machine.sh: status=NO_SOUND sound=0 light=0 t-t0=6
```

```
state_machine.sh: user=393492432127 msg="washing machine has finished!"
```

```
WARNING:yowsup.stacks.yowstack:Implicit declaration of parallel layers in
a tuple is deprecated, pass a YowParallelLayer instead
```

```
INFO:yowsup.demos.sendclient.layer:Message sent
```

`Yowsdown`

`state_machine.sh: new-status=DONE`

`state_machine.sh: old-status=DONE`

Okay, when the timeout expires when we are in the NO_SOUND state, we switch to the DONE state to signal that the washing machine has finished its job:

`state_machine.sh: status=DONE sound=0 light=0 t-t0=1`

`...`

`state_machine.sh: status=DONE sound=0 light=1 t-t0=10`

`state_machine.sh: new-status=LIGHT`

`state_machine.sh: old-status=LIGHT`

Now, I have uncovered the light sensor to simulate that someone has turned on the light in the laundry room:

`state_machine.sh: status=LIGHT sound=0 light=1 t-t0=1`

`...`

`state_machine.sh: status=LIGHT sound=0 light=1 t-t0=6`

`state_machine.sh: new-status=ROOM`

`state_machine.sh: old-status=ROOM`

Again, the timeout has expired, so we can consider that the light has been on for a long time, which means that the user has received our WhatsApp message and they have come into the laundry room to pick up the laundry:

`state_machine.sh: status=ROOM sound=0 light=1 t-t0=1`

`...`

`state_machine.sh: status=ROOM sound=0 light=0 t-t0=8`

`state_machine.sh: new-status=NO_LIGHT`

`state_machine.sh: old-status=NO_LIGHT`

Now, I have covered the light sensor again to simulate that the light in the laundry room has been turned off:

`state_machine.sh: status=NO_LIGHT sound=0 light=0 t-t0=1`
`...`
`state_machine.sh: status=NO_LIGHT sound=0 light=0 t-t0=6`
`state_machine.sh: new-status=IDLE`
`state_machine.sh: old-status=IDLE`
`state_machine.sh: status=IDLE sound=0 light=0 t-t0=1`
`state_machine.sh: new-status=IDLE`
`state_machine.sh: old-status=IDLE`

In the end, after `timeout` has expired, we can return to the `IDLE` state waiting for a new cycle to begin.

The following is the screenshot of my smartphone showing the WhatsApp activity:

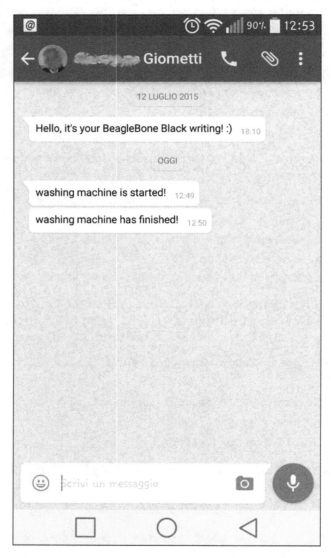

Summary

In this chapter, we discovered how to detect sound and light levels by using specific sensors and how to write a simple Bash script to implement a state machine to manage our laundry room. Also, we discovered how to send some alerting messages to a smartphone through the WhatsApp service.

In the next chapter, we'll try to implement a baby room sentinel to control what happens to our little baby! We'll be able to monitor the room temperature, detect if the baby is crying or if she is actually breathing, and much more.

6

Baby Room Sentinel

In this chapter, we're going to show a possible implementation of a baby room sentinel capable of monitoring the room by detecting whether our baby is crying or if the baby is breathing during sleep. Also, as a special feature, the system will be able to measure the baby's temperature with a contactless temperature sensor.

We'll see several kinds of sensor, such as pressure, sound, and temperature. Also, regarding the temperature sensors, we'll see an interesting infrared version that is useful to measure surface temperature without touching it. Additionally, we'll provide our prototype of a nice, tiny LCD screen in order to see what's going on in the baby's room.

The basics of functioning

When we have a baby, it's quite normal to buy different devices to check when the baby cries or has a fever, or if the baby is still breathing during sleep. So, in this chapter, we'll try to implement several smart sensors to detect these states of danger using our BeagleBone Black and some special sensors.

> Warning! *Let me remind that this project is a prototype and it cannot be used as a personal safety application!* It's simply a study of a possible implementation of a baby room sentinel device.
>
> *Neither the author of this book nor Packt Publishing recommends or endorses that this product be used alone or as a component in any personal safety applications.* The reader is warned about the fact that these sensors and controls do not include the self-checking redundant circuitry needed for such use.
>
> *Neither the author of this book nor Packt Publishing will be held liable for unauthorized use of this prototype. The users can use this device at their own risk!*

To detect when the baby is crying, we can use a sound detector, as we did in the previous chapter; but this time, we should elaborate the input signal a bit more in order to effectively detect whether the baby is really crying or not. By looking at the following screenshot, we can see a simple 40 seconds plot of an audio signal of a crying baby (sample time is *Ts=0.01 s=10ms*):

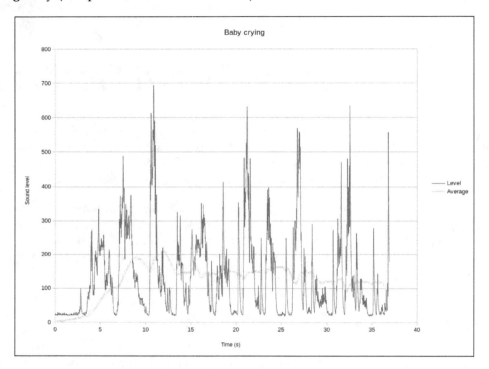

In red is the raw audio signal, while in yellow is the average of that signal over a 5 second window over the current time, that is, the yellow signal is the average value of all collected audio samples in the last 5 seconds.

As already stated, the sampling time *Ts* is *10ms*, which is not suitable for an audio recording but is enough for our purposes. In fact, we can see that by using the average value represented by the yellow line, we can detect if the baby is crying by just using a suitable threshold.

Regarding the breathing, the problem is quite similar; in fact, we can suppose that, more or less, normal breathing during sleep may vary from 12 to 16 breaths per minute, that is, a frequency range from 0.26 Hz to 0.2 Hz. However, this time the average level of the signal is not useful, but we can use its amplitude in a suitable timing window instead. To better explain the concept, consider the following screenshot:

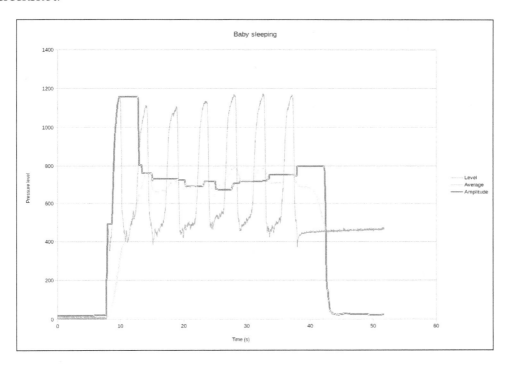

The red color is the raw pressure signal, while the yellow color is the average of that signal over a 5 second window up to the current time. As stated previously, the yellow signal is just the average value of all collected audio samples in the last 5 seconds (sampling time is still $Ts=10\ ms$). The blue color is the amplitude of the pressure signal computed as the difference between the maximum and the minimum value found in the considered temporal window, that is, this time, we find the **maximum value (Vmax)** and the **minimum value (Vmin)** of the collected audio samples in the last 5 seconds each time, and we compute the difference between Vmax and Vmin.

In the first 8 seconds of the preceding plot, the output is near to 0 since there is nothing on the sensor. Then, around $T = 8$ seconds, a baby has been put on the sensor, so it starts to return higher values, as expected. In this situation, both the average and the amplitude of the input signal have grown and they reach (more or less) some stable values. However, the important thing to highlight is when the baby stops breathing (Don't worry! No baby stopped breathing during this test! It's just a simulation.). When the baby stops breathing (this happened near T = 38 seconds), the pressure sensor still detects something, and both the average and the amplitude decrease; but it's the pressure's amplitude that does the biggest jump! As you can see in the preceding screenshot, while the average moves from **700** to **450**, the amplitude moves from **700** to **10**.

For our prototype, we can use the pressure amplitude to detect the baby's breath, and the pressure average to detect if a baby is present or not. The user should notice that both average and amplitude calculations can be done at the same time by using a C function like the following one:

```c
void extract(int arr[], size_t n, int *avg, int *min, int *max)
{
        int i;
        float sum = 0;

        if (min)
                *min = 4096;
        if (max)
                *max = 0;
        for (i = 0; i < n; i++) {
                sum += ((float) arr[i]) / ((float) n);
                if (min)
                        *min = min(*min, arr[i]);
                if (max)
                        *max = max(*max, arr[i]);
        }
        *avg = (int) sum;
}
```

The `extract()` function gets the `arr` array holding the pressure data and, by using a single `for` loop, it can do both calculations in parallel.

Regarding the pressure sensor, we have to take into account that it cannot work well if it is not properly put into a box with a special mechanism suitable to detect breath. In the following screenshot, I show a possible implementation of such a box:

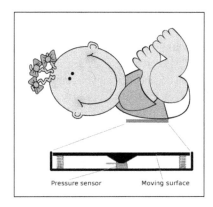

The box should be put under the baby near the back, and attention should be paid to ensure that the upper side is in the right position to capture the movement of the lungs. The top of the box (moving surface) can move up and down, thanks to the springs, and it can detect pressure due to the movement of the baby's lungs with the pin that impinges on the pressure sensor, conveying the pressure.

Now, the last thing to take care of is the digital thermometer to measure the body temperature level. For this purpose, we can use a normal temperature sensor, but since we're talking about babies, we'd like to use a contactless temperature sensor. These special sensors are capable to measure an object's temperature without touching it by using the infrared rays that a surface emits and that are in the *field-of-view* of the sensor. So, when there is nothing in front of the sensor, we can detect the environmental temperature, but when we approach a surface, we can detect the temperature of that surface without actually touching it!

The following screenshot shows a suitable zone to aim the sensor at in order to measure the baby's temperature:

Setting up the hardware

In this project, we are going to use two analog sensors, a digital sensor, and a tiny LCD to implement a little GUI. The analog sensors are connected to two different ADCs, while the digital sensor (the contactless temperature sensor) uses an I²C bus to communicate with the BeagleBone Black. Lastly, the tiny LCD is connected to our BeagleBone Black board by an SPI bus and some GPIOs.

Regarding the alarm devices to alert the parents, we can use a normal buzzer or a more sophisticated SMS gateway, or both. But in any case, the connections of these devices can be retrieved from the preceding chapters, so, due to lack of space, I'm not going to add any of them in this chapter. The reader can try to implement both the hardware and software by themselves as an exercise.

Setting up the contactless temperature sensor

The contactless temperature sensor used in this prototype is shown in the following screenshot:

The devices can be purchased from http://www.cosino.io/product/contactless-temperature-sensor, or found by surfing the Internet.

The user guide of this device is available at https://www.sparkfun.com/datasheets/Sensors/Temperature/SEN-09570-datasheet-3901090614M005.pdf.

This device is really interesting since it's capable of measuring the temperature of an object without touching it! In reality, it is an infrared thermometer with a 17-bit resolution in wide temperature ranges: –40°C to 85°C for ambient temperature and –70°C to 382.2°C for an object's temperature.

The measured value is the average temperature of all objects in the *field-of-view* of the sensor, so it's quite obvious that we can use it to measure the environmental temperature as well as body temperature. We simply need to place the sensor near our body, and there we have it!.

Another important feature of this sensor is that it is a digital device, that is, data can be retrieved by using a digital connection, which is immune to disturbances from the environment, even over (relatively) long distances. So, we can consider to put it on a handpiece for more practical usage.

The bus available for this device is the I²C bus, and the necessary connections are reported in the following table:

Pin	Temperature sensor pin
P9.4 - **VCC**	3 - **VDD**
P9.17 - **SCL**	1 - **SCL**
P9.18 - **SDA**	2 - **SDA**
P9.2 - **GND**	3 - **Vss**

For completeness, the device's pins mapping is shown in the following diagram:

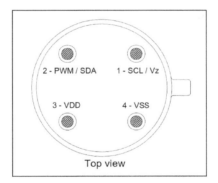

Now, if everything has been properly connected, we can activate the I²C bus with the following command:

```
root@beaglebone:~# echo BB-I2C1 > /sys/devices/bone_capemgr.9/slots
```

Then, by using the i2cdetect command, we should get something as follows:

```
root@beaglebone:~# i2cdetect -y -r 2
     0  1  2  3  4  5  6  7  8  9  a  b  c  d  e  f
00:          -- -- -- -- -- -- -- -- -- -- -- -- --
10: -- -- -- -- -- -- -- -- -- -- -- -- -- -- -- --
```

```
20: -- -- -- -- -- -- -- -- -- -- -- -- -- -- -- --
30: -- -- -- -- -- -- -- -- -- -- -- -- -- -- -- --
40: -- -- -- -- -- -- -- -- -- -- -- -- -- -- -- --
50: -- -- -- -- -- -- -- -- -- -- 5a -- -- -- -- --
60: -- -- -- -- -- -- -- -- -- -- -- -- -- -- -- --
70: -- -- -- -- -- -- -- --
```

Here, we can see that a device at address `0x5a` has answered.

 Note that you may get a different address. In this case, all the following commands must be modified accordingly.

By looking at the datasheet, we discover that the temperature can be retrieved by reading at the device location `0x07`. So, by using the `i2cget` command, we can do the following:

```
root@beaglebone:~# i2cget -y 2 0x5a 0x07 wp
0x3bab
```

The output value can now be converted in **degrees Celsius** (°C) by multiplying it by 0.02 after converting it in a decimal value. So, we can use the following command:

```
root@beaglebone:~# echo "$(printf "ibase=16; %X\n" $(i2cget -y 2 0x5a
0x07 wp) | bc) * 0.02 - 273.15" | bc
32.11
```

To better understand what we do with the preceding command, let me explain it by using a list of equivalent (and more readable) commands starting with the following one where, by using `i2cget`, we get the data from the sensor and store it in the `v_hex` variable:

```
    v_hex=$(i2cget -y 2 0x5a 0x07 wp)
```

Then, we convert the hexadecimal value to a decimal one and store it in the `v_dec` variable by using the `bc` command as follows:

```
    v_dec=$(printf "ibase=16; %X\n" $v_hex | bc)
```

In the end, we simply multiply the decimal value held in the `v_dec` variable by 0.02 to get the temperature in **degrees Kelvin** (°K). Then, we subtract the value 273.15 to get it in °C:

```
    echo "$v_dec * 0.02 - 273.15" | bc
```

Now, to measure body temperature, we simply need to aim the sensor at our head, near the temple, and execute the following command. I get the following output:

```
root@beaglebone:~# echo "$(printf "ibase=16; %X\n" $(i2cget -y 2 0x5a
0x07 wp) | bc) * 0.02 - 273.15" | bc
34.97
```

Great, I'm not ill!

> The reader can take a look at the book *BeagleBone Essentials*, *Packt Publishing*, written by the author of this book, in order to get more information regarding how to activate and use the I²C buses available on the system.

Setting up the pressure sensor

A pressure sensor is shown in the following image:

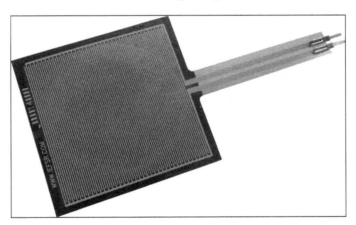

> The devices can be purchased from http://www.cosino.io/product/pressure-sensor, or by surfing the Internet.
>
> The user guide of this device is available at https://www.pololu.com/file/download/fsr_datasheet.pdf?file_id=0J383.

This device can detect (and measure) a force acting on its active surface. In simple words, it can report pressure intensity by varying its internal resistance. From the datasheet, we can see that this resistance may vary from over 1 MΩ, when no force is present, to few hundred Ohms when a force is applied.

By remembering that the BeagleBone Black's ADC inputs must be limited to 1.8V, we can use the circuitry shown in the following diagram to safely read from this sensor (see *Chapter 2, Ultrasonic Parking Assistant*):

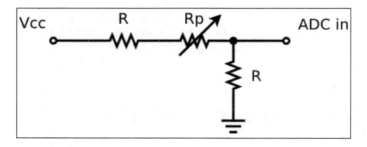

In the preceding diagram, *R=6.8 KΩ* and **Rp** are the pressure sensor's internal resistors, which are represented by a variable resistor.

Note that the preceding circuitry not only connects the sensor to the BeagleBone Black's ADC input pin, but also prevents the input voltage on that pin from going lower than the critical value of 1.8V! The V_{ADCin} voltage is given by the following formula:

$$V_{ADCin} = R / (R + Rp + R) * Vcc = R / (2R + Rp) * Vcc$$

Now, we know that *Vcc* is 3.3V, so, even in case that the *Rp* value drops to 0 Ω, the V_{ADCin} is equal to Vcc/2, that is, 1.65V, which is a safe value for the BeagleBone Black's ADCs.

This sensor must be connected to the BeagleBone Black at the *AIN1* input pin, which is labeled *P9.40*, while the other end must be connected to the resistor R, as shown in the preceding circuitry diagram.

Now, to check all connections, we can enable the BeagleBone Black's ADCs by using the following command:

```
root@beaglebone:~# echo cape-bone-iio > /sys/devices/bone_capemgr.9/slots
```

These settings can be done by using the `bin/load_firmware.sh` script in the book's example code repository, as follows:

```
root@beaglebone:~# ./load_firmware.sh adc
```

Then, we can read the pressure on the sensor with the following command:

```
root@beaglebone:~# cat /sys/devices/ocp.3/helper.12/AIN1
2
```

The preceding value is due to the fact that there is nothing on the sensor; but if we simply try to put a finger on it and then reread the sensor, we get the following result:

```
root@beaglebone:~# cat /sys/devices/ocp.3/helper.12/AIN1
982
```

So, the higher the pressure on the sensor, the higher the returned value.

Setting up the sound detector

The sound detector is the same as the one used in *Chapter 5, WhatsApp Laundry Room Monitor*, so you can take a look at the *Setting up the hardware* section in the same chapter to see how to set up and test this device. However, for the sake of completeness, some basic information about it is provided again, and it's shown in the following image:

The devices can be purchased from `http://www.cosino.io/product/sound-detector`, or by surfing the Internet.

The board is based on the amplifier LMV324, with the datasheet available at `http://dlnmh9ip6v2uc.cloudfront.net/datasheets/Sensors/Sound/LMV324.pdf`, while the board's schematic is available at `http://dlnmh9ip6v2uc.cloudfront.net/datasheets/Sensors/Sound/sound-detector.pdf`.

The connections are in the following table:

Pin	Sound sensor
P9.4 - VCC	VCC
P9.39 - AIN0	R @ENVELOPE
P9.3 - GND	GND

Remember that the ADCs input must be limited to 1.8V, so we must scale the sensor's output voltage by a factor of two, as described in the previous chapter.

Now, to check all connections, we can use the following command:

```
root@beaglebone:~# cat /sys/devices/ocp.3/helper.12/AIN0
24
```

If you try to speak while you rerun the command, you should get a higher value, as follows:

```
root@beaglebone:~# cat /sys/devices/ocp.3/helper.12/AIN0
201
```

So, the higher the environmental sound, the higher the returned value.

Connecting the tiny LCD

The tiny LCD used in this chapter is shown in the following image:

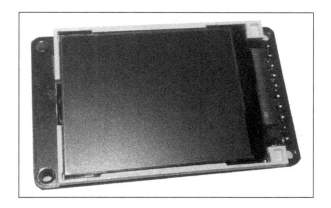

 The device can be purchased from `http://www.cosino.io/` `product/color-tft-lcd-1-8-160x128`, or by surfing the Internet. The LCD is based on the ST7735R chip, which has its datasheet at `https://www.adafruit.com/datasheets/ST7735R_V0.2.pdf`.

As stated previously, to connect the LCD, we must use an SPI bus and some GPIOs available in the BeagleBone Black's expansion connectors. The following table shows the electrical connections between the BeagleBone Black's pins and the LCD pins:

Pin	LCD pin
P9.4 - Vcc	9 - Vcc
P9.29 - MISO	Not connected
P9.30 - MOSI	4 - MOSI
P9.31 - SCLK	3 - SCK
P9.28 - SS0	5 - TFT_CS
P9.25	7 – D/C
P9.27	8 - RESET
P8.19	1 - LITE
P9.2 - GND	10 - GND

To enable the device, we can use a driver that should already be available on your system. To verify it, just use the following command:

```
root@beaglebone:~# zcat /proc/config.gz | grep -i st7735
CONFIG_FB_ST7735=y
```

In my kernel configuration, the driver is statically linked into the kernel, but it's okay to have it as a module. In this case, the output should be something like the following output:

```
CONFIG_FB_ST7735=m
```

After checking the driver, we also need a proper DTS file to set up the kernel. Instead of writing a new one from scratch, I got a suitable DTS file from the following URL by using the wget command:

```
root@beaglebone:~# wget https://raw.githubusercontent.com/beagleboard/
devicetree-source/master/arch/arm/boot/dts/cape-bone-adafruit-lcd-00A0.
dts
```

After the download, we need only to compile the preceding DTS file by using the following command:

```
root@beaglebone:~# dtc -O dtb -o /lib/firmware/cape-bone-lcd-00A0.dtbo -b
0 -@ cape-bone-adafruit-lcd-00A0.dts
```

Now, we can enable the LCD by using the usual echo command:

```
root@beaglebone:~# echo cape-bone-lcd > /sys/devices/bone_capemgr.9/slots
```

> If we get the following error then we have to disable the HDMI support:
>
> ```
> -bash: echo: write error: File exists
> ```
>
> This can be done by editing the u-boot settings in the /boot/uboot/uEnv.txt file and then enabling the following line by uncommenting it:
>
> ```
> optargs=capemgr.disable_partno=BB-BONELT-HDMI,BB-BONELT-
> HDMIN
> ```
>
> Note that on some BeagleBone Black versions, you may find the uEnv.txt file under the /boot directory instead, and the u-boot setting to modify it is as follows:
>
> ```
> cape_disable=capemgr.disable_partno=BB-BONELT-HDMI,BB-
> BONELT-HDMIN
> ```
>
> Then, we only have to reboot the system. Now, if everything is done correctly, we should be able to execute the preceding command without errors.

If everything works well, the BeagleBone Black should enable a colored framebuffer device 32 x 26 characters wide represented in the user space by the **/dev/fb0** device.

The reader can take a look at the book *BeagleBone Essentials, Packt Publishing*, written by the author of this book, in order to get more information regarding how to activate and use the SPI buses available on the system, how to recompile a kernel driver, and to have a brief description of the DTS file.

As a final note, the reader should remember that we can print strings on the LCD by using the following command:

```
root@beaglebone:~# echo "Testing string" > /dev/tty0
```

Here, the /dev/tty0 device is the one connected to the terminal running on the /dev/fb0 framebuffer.

The final picture

The following image shows the prototype I made to implement this project and test the software:

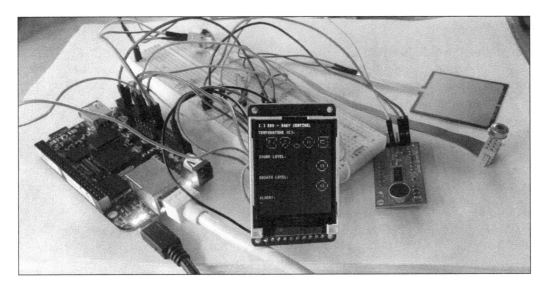

Note that the contactless temperature sensor has been connected to the board by using a flat cable in order to easily move it to measure the temperature of different objects.

Setting up the software

In this project, we're going to show a trick to exchange data between two processes in a very simple manner. At the beginning of the chapter, it was mentioned that the ADCs must be sampled at 100Hz, but we don't need to be so fast to render a simple interface on the external LCD. In fact, a reasonable updating frequency for the user interface can be 1Hz (once per second.) So, to keep the code simple, we implement our device by using two different processes running at different frequencies that exchange data with each other instead of using a single process.

Simply speaking, if we realize a program called adc that reads the data from the ADCs at 100Hz and then prints its output on the stdout stream (standard output) at 1Hz, we can redirect such output to another program called lcd.sh that reads the data from its stdin stream (standard input) at 1Hz and then draws the user interface accordingly.

The data flow is unidirectional. Program adc reads data from the ADC and, after its elaboration stage, sends its output to lcd.sh that manages the LCD. This special functioning is very well represented by a Unix *pipe* as follows:

```
root@beaglebone:~# ./adc | ./lcd.sh
```

The reader should also notice that the timing for the interface is generated by the adc program simply printing its output at well-defined intervals, without any other timing mechanisms, into the lcd.sh program. So, let's see how we can do that.

The ADC manager

As stated previously, to correctly manage and analyze the ADC's input signals, we need a low jitter and fine-grained sampling time. It has already been mentioned that having *Ts=10 ms* as the sampling time would be enough for our purposes, so let's see how we can get it!

In *Chapter 5, WhatsApp Laundry Room Monitor*, we used a simple Bash script to read from the ADC; but for that prototype, the signal frequency was so low that the implemented solution was really reliable. Now we have to do something more elaborate. This time, we're going to use a C program to read data from the ADCs, as shown in the following code snippet:

```c
#define SYSFS_PRESSURE   "/sys/devices/ocp.3/helper.12/AIN1"
#define HZ               100
#define DELAY_US         (1000000 / HZ)

    /* Start sampling the ADC */
```

```
while (1) {
    ret = clock_gettime(CLOCK_MONOTONIC_RAW, &t0);
    EXIT_ON(ret < 0);

    /* Read the ADC */
    fd = open(SYSFS_PRESSURE, O_RDONLY);
    EXIT_ON(fd < 0);
    ret = read(fd, ch, 5);
    EXIT_ON(ret < 1);
    close(fd);
    ret = sscanf(ch, "%d", &val);
    EXIT_ON(ret != 1);

    printf("%ld.%06ld %d\n", t0.tv_sec, t0.tv_nsec / 1000, val);

    /* Calculate the delay to sleep to the next period */
    ret = clock_gettime(CLOCK_MONOTONIC_RAW, &t);
    EXIT_ON(ret < 0);
    delay_us = DELAY_US - difftime_us(&t0, &t);
    EXIT_ON(delay_us < 0);
    usleep(delay_us);
}
```

The complete code can be found in the chapter_06/adc_simple.c file in the book's example code repository.

The code can be compiled directly on the BeagleBone Black by using the make command.

The code functioning is simple. First, we get the current time with the clock_gettime() function. Then, we read the data from the ADC by accessing it from the sysfs interface. And, at the end, we compute the amount of time to sleep for before reaching the new active period.

By running the preceding code, we get the following output:

```
root@beaglebone:~/chapter_06# ./adc_simple
317330.142227 0
317330.153381 10
317330.163604 7
317330.174134 10
317330.184298 5
317330.194473 10
317330.204696 7
```

```
317330.214955 7
317330.225119 13
317330.235331 10
317330.245558 1
317330.255714 10
317330.265858 10
317330.276034 10
317330.286186 7
317330.296346 7
317330.306500 9
317330.316646 8
317330.326924 0
...
```

As we notice from the preceding output, the program is quite precise; but if we use a simple awk script to compute the minimum, maximum, and average jitter value over 1,000 samples, we discover that the program is not so precise:

```
root@beaglebone:~# ./adc_simple | awk -v T=0.01 -v N=1000 -f jitter.awk
avg=0.000255 min=0.000078 max=0.012252
```

> The awk script file jitter.awk can be found in the chapter_06/jitter.awk file in the book's example code repository.

The average and the minimum values are acceptable, but the maximum one is really high. Moreover, sometimes the following may happen:

```
root@beaglebone:~/chapter_06# ./adc_simple
319111.158747 0
319111.168981 8
319111.179131 9
319111.189269 10
319111.199439 10
319111.209586 11

...

319113.140526 0
adc_simple.c[  65]: main: fatal error in main() at line 65
```

This error happens when the delay computed in the following lines becomes negative:

```
delay_us = DELAY_US - difftime_us(&t0, &t);
EXIT_ON(delay_us < 0);
```

If the system is too slow in scheduling the process at the beginning of the period, it may happen that we do not have enough time to complete our task! This is due to the fact that we are not using a real-time system and we have no guarantees of correct scheduling.

However, we can try to resolve this problem with some tricks. The Linux kernel is not real time, but it has some facilities that may help us in realizing an acceptable compromise. In fact, the system allows us to use different schedulers to manage the processes running on the BeagleBone Black. In particular, we can use the `chrt` command to manipulate the real-time schedule attributes of a process and then to set the **FIFO scheduler** that may help us to reduce the `jitter` value and the scheduling delay error. If we rerun the preceding test using the `chrt` command as follows, we get a different result:

```
root@beaglebone:~# chrt -f 99 ./adc_simple | awk -v T=0.01 -v N=1000 -f jitter.awk
avg=0.000102 min=0.000022 max=0.000781
```

Also, the scheduling delay error disappears!

Note that even using the `chrt` command, the Linux kernel is not real time anyway, so nobody can guarantee that everything will go well forever! To make the system reliable, we have to add some recovery code in case something goes wrong.

Considering what was just explained, a possible implementation of the ADC manager is reported in the following code snippet:

```
/* Set stdout line buffered */
setlinebuf(stdout);

/* Do a dummy read to init the data buffers */
c = read_adc(SYSFS_SOUND);
for (snd_idx = 0; snd_idx < ARRAY_SIZE(snd); snd_idx++)
    snd[snd_idx] = c;
c = read_adc(SYSFS_PRESSURE);
for (prs_idx = 0; prs_idx < ARRAY_SIZE(prs); prs_idx++)
    prs[prs_idx] = c;
```

```
/* Set FIFO scheduling */
param.sched_priority = 99;
ret = sched_setscheduler(getpid(), SCHED_FIFO, &param);
EXIT_ON(ret < 0);

/* Start sampling the ADC */
snd_idx = prs_idx = 0;
ret = clock_gettime(CLOCK_MONOTONIC_RAW, &t);
EXIT_ON(ret < 0);
while (1) {
    ret = clock_gettime(CLOCK_MONOTONIC_RAW, &t0);
    EXIT_ON(ret < 0);

    /* Read the data from the ADCs */
    snd[snd_idx] = read_adc(SYSFS_SOUND);
    prs[prs_idx] = read_adc(SYSFS_PRESSURE);

    /* Extract informations from buffered data */
    extract(snd, ARRAY_SIZE(snd),
        &snd_avg, NULL, NULL);
    extract(prs, ARRAY_SIZE(prs),
        &prs_avg, &prs_min, &prs_max);
    dbg("%ld.%06ld prs:%d min=%d max=%d snd:%d",
        t0.tv_sec, t0.tv_nsec / 1000,
        prs[prs_idx], prs_min, prs_max, snd[snd_idx]);

    /* We have to output the pressure data each second,
     * that is every HZ ticks.
     * Also we have to read the sound level...
     */
    if (ticks++ == 0)
        printf("%d %d %d\n",
            prs_avg, prs_max - prs_min, snd_avg);
        ticks %= HZ;

        /* Calculate the delay to sleep to the next period */
        ret = clock_gettime(CLOCK_MONOTONIC_RAW, &t);
        EXIT_ON(ret < 0);
        delay_us = DELAY_US - difftime_us(&t0, &t);
        EXIT_ON(delay_us < 0);
        usleep(delay_us);

        /* Move the index */
        prs_idx++;
        prs_idx %= ARRAY_SIZE(prs);
```

```
        snd_idx++;
        snd_idx %= ARRAY_SIZE(snd);
    }
```

 The complete code can be found in the `chapter_06/adc.c` file in the book's example code repository.

The `setlinebuf()` function is needed to force an output at each printed line, while the `sched_setscheduler()` function is used to enable the FIFO scheduler (as the `chrt` command does). The code is quite similar to before except the fact that we use the `extract()` function (mentioned at the beginning of this chapter) to calculate the average, minimum, and maximum values of the input data as requested. Note that the program prints its output once per second, thanks to the `ticks` variable.

If executed, the program will then print several lines, one per second, reporting the pressure average value, the pressure signal amplitude, and the sound average value, as follows:

```
root@beaglebone:~# ./adc
0  16  20
0  19  21
1  21  21
2  22  23
3  22  23
4  22  24
...
```

All this data is taken as inputs by the `lcd.sh` process described in the following section.

 Note that the BeagleBone Black's ADCs have a continuous mode function that can be used to reach a higher sampling rate, but I didn't use it in this project due to the fact that it's not strictly needed nor supported on all kernels.

The curious reader can get further information about this topic at `http://processors.wiki.ti.com/index.php/AM335x_ADC_Driver's_Guide`.

The LCD manager

The program that manages the LCD is a simple Bash script that uses some tricks to realize a fancy rendering of the collected data.

As stated previously, this program runs each second, thanks to the ADC manager that sends its output periodically to the `lcd.sh` program. A simple functioning of this program can be represented by the following meta-code:

```
while true ; do
    wait_for_data_from_ADC
    render_data_to_LCD
done
```

That's all! The other complexities are only related to how we wish to implement the user interface.

Regarding this issue, I decided to use a really simple solution: some terminal **escape sequences** to manage the colors, and the `figlet` program to *draw* big fonts. Escape sequences are used to easily print some characters on the screen with specified colors by using the `echo` command, as follows:

```
root@beaglebone:~# echo -e "\e[31mRED TEXT\e[39m"
RED TEXT
```

The `\e[31m` sequence sets the red color, while the `\e[39m` sequence resets the default one.

> For further information regarding these sequences, a good starting point is available at https://en.wikipedia.org/wiki/ANSI_escape_code.

The `figlet` program is a tool that can be used to simulate printing big fonts on a terminal, a kind of ASCII art. To install it, we can use the following command:

```
root@beaglebone:~# aptitude install figlet
```

Then, its usage is very simple, as shown in the following example:

```
root@beaglebone:~# figlet "simple string"
```

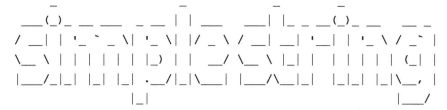

For our user interface implementation, I used some special option arguments that I'm not going to explain here due to lack of space, but a curious reader can take a look at man pages of `figlet` for further information.

After this brief introduction, it's time to show a snippet of the main code of the `lcd.sh` program:

```
# Ok, do the job
clear_scr

tick=1
while true ; do
    # Read the temperature from the sensor and convert it in C
    t=$(i2cget -y 2 0x5a 0x07 wp)
    t=$(hex2dec $t)
    t=$(echo "$t * 0.02 - 273.15" | bc)

    # Read the pressure and sound data from the " adc" tool
    read -u 0 v b s

    # Draw the GUI

    # Check for a minimum pressure, otherwise drop to 0 sound and
    # pressure data in order to not enable any alarm
    if [ $v -lt $PRS_AVG ] ; then
        s=0
        b=0
        enabled="false"
    else
        enabled="true"
    fi

    # Rewrite the screen
    goto_xy 0 0

    echo -en "[${CH_PULSE:$tick:1}] "
```

```
    echo -e "${FC_LIGHT_MAGENTA}BBB - BABY SENTINEL${FC_DEFAULT}\n"

    echo -en "TEMPERATURE (C):"
    if (( $(bc <<< "$t > 37.00") == 1 )) ; then
        echo -e "$FC_RED"
        t_alrm="true"
    else
        echo -e "$FC_GREEN"
        t_alrm="false"
    fi
    figlet -f small -W -r -w 32 "$t"
    echo -e "$FC_DEFAULT"

    echo -en "SOUND LEVEL:"
    if $enabled && [ $s -gt $SND_AVG ] ; then
        echo -e "$FC_RED"
        s_alrm="true"
    else
        echo -e "$FC_DEFAULT"
        s_alrm="false"
    fi
    figlet -f small -W -r -w 32 "$s"

    echo -en "BREATH LEVEL:"
    if $enabled && [ $b -lt $PRS_AMP ] ; then
        echo -e "$FC_RED"
        b_alrm="true"
    else
        echo -e "$FC_DEFAULT"
        b_alrm="false"
    fi
    figlet -f small -W -r -w 32 "$b"
    echo -en "${ES_CLEAR_LN}${FC_LIGHT_RED}ALARMS: ${FC_DEFAULT}"
    $t_alrm && echo -en "${BC_RED}TEMP. "
    $s_alrm && echo -en "${BC_RED}SOUND "
    $b_alrm && echo -en "${BC_RED}BREATH "
    echo -e "${BC_DEFAULT}"

    # Print some debugging messages if requested
    dbg "$(printf "t=%0.2f v=% 4d b=% 4d s=% 4d" $t $v $b $s)"
    dbg "PRS_AVG=$PRS_AVG PRS_AMP=$PRS_AMP SND_AVG=$SND_AVG"

    tick=$(( ($tick + 1) % ${#CH_PULSE} ))
done
```

The complete code can be found in the `chapter_06/lcd.sh` file in the book's example code repository.

From the beginning of the `while` loop until the `read` statement, we simply collect the data. Then, the code following the `Draw the GUI` comment is just to render the user interface. Note that the line with the `read` command will wait until an input line arrives from the `stdin` stream, that is, from the `adc` program.

To test it via SSH in a normal terminal of our host system, simulating the LCD, we can execute the following command to reduce the size of the terminal's windows to 32x26 characters, which is the size of the terminal on LCD:

```
root@beaglebone:~# echo -e '\e[8;26;32t'
```

 Note that the preceding command is just another escape sequence.

Then, we can execute the program, as follows:

```
root@beaglebone:~# ./adc | ./lcd.sh
```

The output is as shown in the following screenshot:

Final test

To test the prototype, I used some tricks to simulate the baby: I got the crying sound on the Internet and simply reproduced it with an audio player. Regarding the breath, I used doll, manually pressurizing its chest in time with my breathing. I admit it's not the best test, but my children are too big to help me in these experiments!

To set up all peripherals and drivers, we can use `SYSINIT.sh`, as in the following command:

```
root@beaglebone:~# ./SYSINIT.sh
done!
```

 This command can be found in the `chapter_06/SYSINIT.sh` file in the book's example code repository.

Then, I executed both the `adc` and `lcd.sh` programs by using the following command line in order to send all outputs to the terminal that runs on the tiny LCD:

```
root@beaglebone:~# ./adc | ./lcd.sh > /dev/tty0
```

 Note that on the first framebuffer device, we have at least one terminal defined by default, which is referred to by the `/dev/tty0` device.

Summary

In this chapter, we discovered a more reliable and precise way to get access to the BeagleBone Black's ADCs and learned how we can get access to an I^2C device by using a raw access to the bus. This was done in order to be able to manage a pressure sensor and a contactless temperature sensor. Also, we discovered how to connect a tiny LCD via the SPI bus to our BeagleBone Black board to add a little user interface.

In the next chapter, we'll try to implement a plant monitor to measure what happens to our beloved plants! Also, we will discover how we can periodically take some pictures and then publish them on a Facebook account.

7
Facebook Plant Monitor

Social networks are very common nowadays, and having a monitoring (or controlling) system that interacts with them has become a must, especially for consumer systems.

In this chapter, we're going to see how to implement a plant monitor capable of measuring solar light, soil moisture, and the soil temperature (internal and external) along with how to take some pictures at specific intervals via a webcam.

The user will be able to control the monitor through a web interface and they can then decide to publish the plant pictures on their Facebook timeline.

The basics of functioning

In this project, I'm going to present a simple implementation of a plant monitor with the following two special features:

- The first feature is the ability to measure and estimate the soil moisture according to a direct measurement of the moisture via a dedicated sensor and through the temperature difference between the external soil temperature and the internal soil temperature in the garden pot where the plant lives. This is because the internal resistance of the moisture sensor may change with the temperature; in fact, when the sun beats down on the soil and the soil warms up, the resistance changes and this effect will produce a false *dry* read. For this reason, we employ two different temperature probes in order to know if the soil is too hot with respect to the internal soil temperature and then we regulate the moisture level.

- The second feature is the ability to add a webcam to take several pictures of our lovely plant at well-defined intervals and the possibility to publish them, at the user's request, on our Facebook timeline in order to show to our friends how green is our thumb!

A little schematic representation of the sensors position to correctly implement the first feature is shown in the following image:

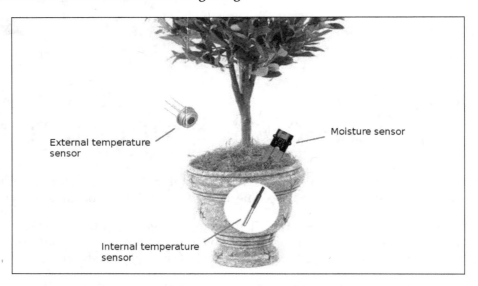

Let's consider Te as the external soil temperature and Ti as the internal soil temperature. If we name the value returned by the moisture sensor M, a reasonable estimation (Me) of the real moisture can be given by the following formula:

$Me = M + K * (Te – Ti)$ *when Te > Ti*

$Me = M$ *when Te ≤ Ti*

where K is a suitable (and empirical) *moisture coefficient* that the user can set at runtime in order to fix their needs. Note that if K is 0, the system will take the measured moisture level without any compensation.

Regarding the second feature, it is possible to take a picture of the monitored plant and then publish it on Facebook. In order to do this, we have to use the Facebook API to interact with a Facebook account. This step will be explained in detail in the next section.

Setting up the hardware

In this project, we are going to use two analog sensors, two digital sensors, and a webcam to take the pictures. The analog sensors are connected to two different ADC channels inputs (as in the previous chapter). The contactless external temperature sensor uses an I²C bus to communicate with the BeagleBone Black, while the waterproof internal temperature sensor uses a 1-Wire bus. Finally, the webcam is connected via the **USB** bus.

As in the previous chapter, I can add some actuators to flood water the plant and so on; but due to lack of space, I decided to leave these tasks to the reader as an exercise.

Connecting the moisture sensor

The moisture sensor is the device shown in the following image:

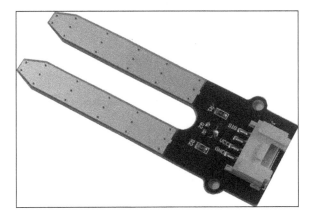

The device can be purchased from http://www.cosino.io/product/moisture-sensor, or by surfing the Internet.
The user guide of this device is available at http://seeedstudio.com/wiki/Grove_-_Moisture_Sensor.

This device is quite similar to the water sensor presented in *Chapter 3, Aquarium Monitor*, since the functioning is still based on water conductivity; however, its shape is different due to the fact that we are interested in the soil humidity level and not whether water is present or not.

The connection to the BeagleBone Black is shown in the following table:

Pin	Cable color
P9.3 - Vcc	Red
P9.39 - AIN0	Yellow
P9.1 - GND	Black

By using a multimeter, we can verify whether the output voltage is near to 0V when the sensor is in the air, the output is about 1.5V while in water, and that all the other output values stay in this range.

Note that, since the maximum output voltage is about 1.5V, we can safely connect the output pin of this device directly to the BeagleBone Black ADC's input. However, we can use a 1.8V Zener diode between the *AIN0* pin and GND in order to be definitely sure that the output voltage cannot go over the critical threshold of 1.8V. (Remember what has been said several times in the previous chapters, starting from *Chapter 2, Ultrasonic Parking Assistant*, about this issue.)

Now, to verify the device output, we can enable the BeagleBone Black's ADCs by using the following command:

```
root@beaglebone:~# echo cape-bone-iio > /sys/devices/bone_capemgr.9/slots
```

These settings can be done by using the `bin/load_firmware.sh` script in the book's example code repository, as follows:

```
root@beaglebone:~# ./load_firmware.sh adc
```

Then, we can read the moisture level in water by putting the sensor into a cup of water and then executing the following command:

```
root@beaglebone:~# cat /sys/devices/ocp.3/helper.12/AIN0
1745
```

Then, we can read the moisture level in air by removing the sensor from the water and then rerunning the following command:

```
root@beaglebone:~# cat /sys/devices/ocp.3/helper.12/AIN0
1
```

So, the higher the moisture level, the higher the returned value.

Connecting the light sensor

An ambient light sensor is shown in the following image:

The device can be purchased from http://www.cosino.io/ product/photoresitor, or found by surfing the Internet.

The user guide of this device is available at https://www.sparkfun.com/datasheets/Sensors/Imaging/SEN-09088-datasheet.pdf.

This device functioning is quite similar to the light sensor presented in *Chapter 5, WhatsApp Laundry Room Monitor*; however, this is a single photoresistor with low power consumption that has an output function very similar to the pressure sensor presented in *Chapter 6, Baby Room Sentinel*, so, even in this case, we can use the same circuitry to manage it, as reported in the following diagram:

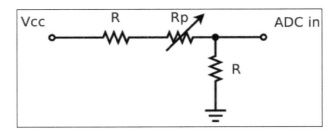

In the preceding diagram, $R=6.8\ K\Omega$ and **Rp** is the light sensor's internal resistor that is represented by a variable resistor.

Note that the preceding circuitry not only connects the sensor to the BeagleBone Black's ADC input pin but also ensures that the input voltage on that pin is lower than the critical value on 1.8V! In fact, the V_{ADCin} voltage is given by the following formula:

- $V_{ADCin} = R / (R + Rp + R) * Vcc = R / (2R + Rp) * Vcc$

Now we know that Vcc is 3.3V, so, even in case the **Rp** value drops to 0Ω, the V_{ADCin} is equal to Vcc/2, that is, 1.65V, which is a safe value for the BeagleBone Black's ADCs.

This sensor must be connected with the BeagleBone Black at the *AIN1* input pin, which is labeled *P9.40*, while the other end must be connected with the resistor R, as shown in the preceding circuitry diagram.

Now for all connections, we can use the following command:

```
root@beaglebone:~# cat /sys/devices/ocp.3/helper.12/AIN1
317
```

Now, as we have already done for the other light sensor in the earlier chapters, we can cover it with a cup and try to reread the ADC:

```
root@beaglebone:~# cat /sys/devices/ocp.3/helper.12/AIN1
15
```

On the other hand, if we turn on a lamp over the sensor, we get the following:

```
root@beaglebone:~# cat /sys/devices/ocp.3/helper.12/AIN1
954
```

So, the brighter the environment light, the higher the returned value.

Setting up the contactless temperature sensor

The contactless temperature sensor is the same as that used in *Chapter 6, Baby Room Sentinel,* so you can take a look at the *Setting up the contactless temperature sensor* section to see how to set up and test this device; however, for completeness, again, some basic information about it is reported as follows:

 The device can be purchased from `http://www.cosino.io/product/contactless-temperature-sensor`, or found by surfing the Internet.

The user guide of this device is available at `https://www.sparkfun.com/datasheets/Sensors/Temperature/SEN-09570-datasheet-3901090614M005.pdf`.

The connections are shown in the following table:

Pin	Temperature sensor pin
P9.4 - Vcc	3 - Vdd
P9.17 - SCL	1 - SCL
P9.18 - SDA	2 - SDA
P9.2 - GND	3 - Vss

Now, if everything has been well connected, we can activate the I²C bus with the following command:

```
root@beaglebone:~# echo BB-I2C1 > /sys/devices/bone_capemgr.9/slots
```

While the temperature can be retrieved by reading at the device location 0x07, so by using the i2cget command, we can do the following:

```
root@beaglebone:~# i2cget -y 2 0x5a 0x07 wp
0x3bab
```

The output value must be converted to degree Celsius by multiplying it by 0.02 after converting it to a decimal value, so we can use the following command:

```
root@beaglebone:~# echo "$(printf "ibase=16; %X\n" $(i2cget -y 2 0x5a
0x07 wp) | bc) * 0.02 - 273.15" | bc
```

Setting up the waterproof temperature sensor

Regarding the internal temperature sensor, I used the same one used in *Chapter 3, Aquarium Monitor*, so you can take a look at the *Connecting the temperature sensor* section to see how to set up and test this device; however, for completeness, again, some basic information about it is reported:

The device can be purchased from `http://www.cosino.io/product/waterproof-temperature-sensor`, or found by surfing the Internet.

The datasheet of this device is available at `http://datasheets.maximintegrated.com/en/ds/DS18B20.pdf`.

The connections are shown in the following table:

Pin	Cable color
P9.4 - Vcc	Red
P8.11 - GPIO1_13	White
P9.2 - GND	Black

To enable it, we have to load a proper DTS file into the kernel with the following command:

```
root@beaglebone:~# echo BB-W1-GPIO > /sys/devices/bone_capemgr.9/slots
```

See *Chapter 3, Aquarium Monitor*, to see how to obtain the DTS file.

If everything works well, we should see a new 1-Wire device under the /sys/bus/w1/devices/ directory as follows:

```
root@beaglebone:~# ls /sys/bus/w1/devices/
28-000004b541e9  w1_bus_master1
```

The new temperature sensor is represented by the directory named 28-000004b541e9 and to read the current temperature, we can use the cat command on the w1_slave file, as follows:

```
root@beaglebone:~# cat /sys/bus/w1/devices/28-000004b541e9/w1_slave
d8 01 00 04 1f ff 08 10 1c : crc=1c YES
d8 01 00 04 1f ff 08 10 1c t=29500
```

In the preceding example, the current temperature is t=29500, which is expressed in **millidegrees Celsius (m°C)**, so it's equivalent to 29.5°C.

 Note that your sensor has a different ID, so, in your system, you'll get a different path name in the form of /sys/bus/w1/devices/28-NNNNNNNNNNNN/w1_slave.

Setting up the webcam

In *Chapter 3, Aquarium Monitor,* I explained how to use a webcam with the BeagleBone Black, so you can take a look at that chapter to do the job. However, in that chapter, I required a UVC-based webcam (or at least another one which is supported by the mjpg-streamer tool). But this time, you can use any webcam supported by the Video4Linux driver class.

 The curious reader can obtain more information about Video4Linux drivers from http://www.linuxtv.org/wiki/index.php/Main_Page.

To know if our webcam is supported by the Video4Linux system, you just need to connect it to the BeagleBone Black USB host port and then check if a new /dev/video**X** device has been added (where **X** can be 0, 1, 2, and so on).

As an example, on my system, using the same webcam used in *Chapter 3, Aquarium Monitor*, I get the following output:

```
root@beaglebone:~# ls -l /dev/video*
crw-rw---T 1 root video 81, 0 Jan  1  2000 /dev/video0
```

So, my webcam is supported by the Video4Linux subsystem and it can be used by the fswebcam program to take pictures. To install the program, we can use the following command:

```
root@beaglebone:~# aptitude install fswebcam
```

Then, we can take a simple picture by using the following command:

```
root@beaglebone:~# fswebcam webcam-shot.jpg
--- Opening /dev/video0...
Trying source module v4l2...
/dev/video0 opened.
No input was specified, using the first.
Adjusting resolution from 384x288 to 352x288.
--- Capturing frame...
Captured frame in 0.00 seconds.
--- Processing captured image...
Writing JPEG image to 'webcam-shot.jpg'.
```

It can happen that when the fswebcam program is executed, you get the following message:

```
--- Opening /dev/video0...
Trying source module v4l2...
/dev/video0 opened.
No input was specified, using the first.
Adjusting resolution from 384x288 to 352x288.
--- Capturing frame...
gd-jpeg: JPEG library reports unrecoverable error: Not
a JPEG file: starts with 0x11 0x80
Captured frame in 0.00 seconds.
--- Processing captured image...
Writing JPEG image to 'webcam-shot.jpg'.
```

Then, when you try to display the picture, the image is black. To resolve this issue, the trick is to skip the first frames with the -S option argument, as follows:

```
root@beaglebone:~# fswebcam -S 10 webcam-shot.jpg
```

However, what is really interesting in this program is the possibility to add text in different positions on the picture. For example, by using the following command, we can take a picture with, in the bottom-right corner, the text `Picture's title`, and in the bottom-left corner the text `Information text` with a smaller font and a time stamp:

```
root@beaglebone:~# fswebcam --title "Picture's title" --info "Information
text" --jpeg 85 webcam-shot.jpg
--- Opening /dev/video0...
Trying source module v4l2...
/dev/video0 opened.
No input was specified, using the first.
Adjusting resolution from 384x288 to 352x288.
--- Capturing frame...
Captured frame in 0.00 seconds.
--- Processing captured image...
Setting title "Picture's title".
Setting info text "Information text".
Setting output format to JPEG, quality 85
Writing JPEG image to 'webcam-shot.jpg'.
```

The output picture is shown in the following screenshot:

As you can see, this is a great software for taking pictures with descriptions!

Adding a water pump

Since we know if water is present on the floor, we can also decide to implement an automatic system to flood our plant with water when we detect that it's missing. A possible solution is using a water pump, as already done in *Chapter 3, Aquarium Monitor*. Then, we can use one or more relays to control several pumps by using BeagleBone Black's GPIOs, but, as already stated, I'm not going to do it here due to lack of space, thus, leaving this issue as a useful exercise for the reader.

The final picture

The following image shows the prototype I made to implement this project and to test the software:

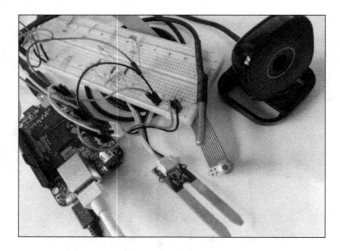

The connections are very simple this time. Note also that all the sensors should be connected to the BeagleBone Black with a long cable in order to be easily put in the plant's garden pot.

Setting up the software

This time, we monitor (and possibly control) a very slow system (a growing plant). So, using a simple Bash script is more than appropriate, while the web interface can be implemented by using two little HTML (with JavaScript) and PHP scripts. The real difficulty, apart from getting access to the peripherals, is using the Facebook API in order to get access to an account.

Apropos the monitoring loop and the web interface, there is a major issue, that is, the **Inter Process Communication (IPC)** system to use in order to exchange data between them. In *Chapter 1, Dangerous Gas Sensors*, we mainly used a MySQL server to store the system's data and configuration settings. But we also used it to exchange data between the different tasks composing the system! However, this time, we have very few data to store and using a database may be very expensive. So, I decided to use a simple way to solve the problem: I'm going to use a file! Yes, by using a normal file with a well-defined internal structure and proper locking functions to serialize access to it, we can solve the problem with very few system resources. This solution will be explained in detail in a later section.

The Facebook API

At `https://developers.facebook.com/docs/reference/php/5.0.0`, there is a user manual of the PHP API I used to get access to my Facebook account.

> Actually, different APIs exist to get access to Facebook, and for further information about them, the reader can take a look at `https://developers.facebook.com/docs/apis-and-sdks`.

There is a lot to read, so in the following section I've listed all steps I followed to install and configure it.

Downloading the code

First of all, we have to download the source code. There are two possible ways to do it: installing with **Composer** and manual installing. I decided to use the manual installing method since I have no Composer support on my BeagleBone Black (nor do I wish to install it for such an easy task).

> The Composer tool is a dependency manager for PHP; a curious reader can refer to `https://getcomposer.org/` for further information about it.

The code can be downloaded by clicking either on the **Download the SDK for PHP v5.0** button in the browser running on the host PC or directly on the BeagleBone Black by using the following command:

```
root@beaglebone:~# wget https://github.com/facebook/facebook-php-sdk-v4/archive/5.0-dev.zip
```

A copy of that archive can be found in the `chapter_07/5.0-dev.zip` file in the book's example code repository.

Once downloaded, put the archive on your BeagleBone Black (if not already present) and then unzip it with the following command:

```
root@beaglebone:~# unzip 5.0-dev.zip
```

If not installed, the `unzip` command can be retrieved by using the usual `aptitude` command.

Now, to use the API, we simply need to use the following two lines in all our scripts:

```
define('FACEBOOK_SDK_V4_SRC_DIR', __DIR__ . '/facebook-php-sdk-v4-5.0-
dev/src/Facebook/');
require_once __DIR__ . '/facebook-php-sdk-v4-5.0-dev/src/Facebook/
autoload.php';
```

To simplify this operation, I created a file called `setup.php`, where I put these lines so that I can do a simple include in all my scripts, as follows:

```
require_once "setup.php";
```

The preceding file can be found in the `chapter_07/setup.php` file in the book's example code repository.

Creating a new Facebook app

Now, we need a new app that can use the API. In fact, all Facebook access must be done through a dedicated app, and the following steps are needed to create a new one:

1. Go to the Facebook developers page at `https://developers.facebook.com/apps/?action=create` and click on the **Add a New App** button.

2. Choose the **Website** option.

3. Type a name for your App (I used `BBB Plant Monitor`) and click on the **Create New Facebook App ID** button.

4. Choose a category (I used `Entertainment`) and click on **Create App ID**.

5. Fill in the **Site URL** with `http://localhost` and click on **Next**.

Note that to be able to follow the preceding steps, you need a pre-existing Facebook account, otherwise the system will ask to register before continuing.

Your App has now been created. Then, in the **Next Steps** section, click on **Skip to Developer Dashboard**, and the panel shown in the following screenshot should appear:

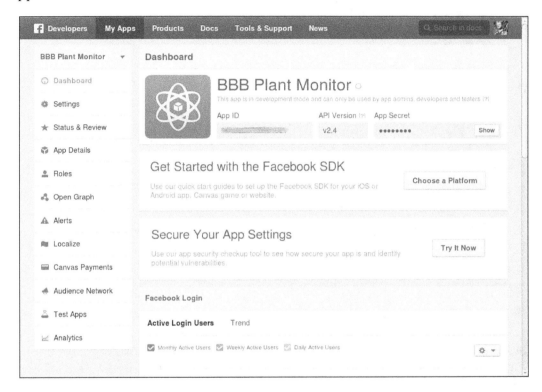

The **App ID** is visible by default, but the **App Secret** is hidden. Click on **Show**, enter your password again (if requested) to reveal the **App Secret**, and then copy and keep this information apart.

Now, to finish the job, click on **Settings**, add http://localhost to **App Domains**, add your valid e-mail to **Contact Email**, and then click on **Save Changes**.

Getting access to the Facebook account

Now that our app is ready, we can start trying some code on it. However, before going live, we can create a `test user` account with different permissions that do not interfere with our real Facebook account to make sure that everything is set up correctly (under the **Roles** menu and then the **Test Users** section).

The following screenshot shows the **Test Users** panel:

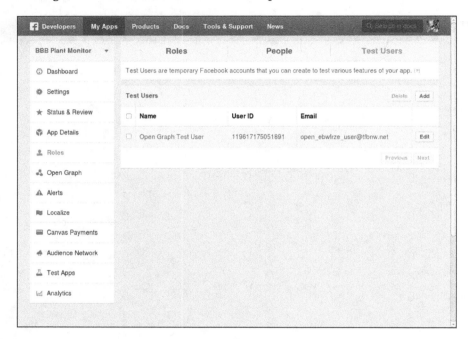

Click on the **Edit** button and choose the **Get an access token for this test user** entry from the displayed list, accept the default settings, and then copy and keep the access token apart.

Now it's time to test our first script! The following code snippet of a simple script used to get basic information of a Facebook account:

```
# Define the Facebook session
$fb = new Facebook\Facebook([
    'app_id'                => APP_ID,
    'app_secret'            => APP_SECRET,
    'default_graph_version' => 'v2.4',
    'default_access_token'  => DEF_TOKEN,
    'fileUpload'            => true,
    'cookie'                => true,
```

```
    ]);

    # Print user's information
    try {
        $res = $fb->get('/me');
    } catch(Exception $e) {
        err("error!\n");
        dbg("=======================================================
==========\n");
        dbg($e);
        dbg("=======================================================
==========\n");
        die();
    }
    $node = $res->getGraphObject();
    info("name is \"%s\" (%s)\n",
        $node->getProperty('name'), $node->getProperty('id'));

    # Print user's permissions
    $res = $fb->get("/me/permissions");
    $node = $res->getDecodedBody();
    info("permissions are:\n");
    foreach ($node['data'] as $perm)
        info("\t%s is %s\n", $perm['permission'], $perm['status']);
```

> The complete code can be found in the chapter_07/get_info.php file in the book's example code repository.

The script is really simple. As the first step, we need to define a new Facebook session with the Facebook\Facebook class, where the APP_ID and APP_SECRET values are taken from the preceding app information, and the DEF_TOKEN is the test that the user's token generated previously. All this information is stored in a dedicated file named config.php, holding the following code:

```php
<?php

define('APP_ID', '000000000000000');
define('APP_SECRET', '000000000000000000000000000000000');
define('DEF_TOKEN', 'XXXXXXXXXXXXXXXXXXXXXXXXXXXXXXXXXXXXXXXX');

?>
```

Please note that all information has been replaced by zeros or the X character, so you have to replace them with your custom values.

Then, the file can be loaded, as in the preceding command, with the following PHP command:

```
require_once "config.php";
```

Once the Facebook session has been created, we can start fetching data from it by using the get() method with different parameters. For example, by using the /me string, we can get the user's name and ID, while using the /me/permissions string, we can get the user's permissions.

 The Facebook permissions are listed and explained well at https://developers.facebook.com/docs/facebook-login/permissions/v2.2.

Once collected, the user's information is displayed as follows:

```
root@beaglebone:~# ./get_info.php
get_info.php[  62]: name is "Open Graph Test User" (119617175051891)
get_info.php[  67]: permissions are:
get_info.php[  69]:     user_friends is granted
get_info.php[  69]:     publish_actions is granted
get_info.php[  69]:     public_profile is granted
```

The API is okay and the accessing information has been correctly written! So, let's see how we can post a picture on the test user's timeline. To do the trick, we must use the post() method, as shown in the following code snippet:

```
# Define the Facebook session
$fb = new Facebook\Facebook([
    'app_id'                => APP_ID,
    'app_secret'            => APP_SECRET,
    'default_graph_version' => 'v2.4',
    'default_access_token'  => DEF_TOKEN,
    'fileUpload'            => true,
    'cookie'                => true,
]);

# Print user's information
try {
```

```
    $res = $fb->get('/me');
} catch(Exception $e) {
    err("error!\n");
    dbg("==========================================================
==========\n");
    dbg($e);
    dbg("==========================================================
==========\n");
    die();
}
$node = $res->getGraphObject();
info("name is \"%s\" (%s)\n",
    $node->getProperty('name'), $node->getProperty('id'));

# Publish to user's timeline
try {
    $ret = $fb->post('/me/photos', array(
        'message'    => 'MyPlant message',
        'source'     => $fb->videoToUpload(realpath('web-cam-shot.jpg')),
    ));
} catch(Exception $e) {
    err("error!\n");
    dbg("==========================================================
==========\n");
    dbg($e);
    dbg("==========================================================
==========\n");
    die();
}

    info("done\n");
```

After creating the Facebook session, as explained previously, we have to call the post() method with the /me/photos string and proper parameters. In particular, you should pay attention to the videoToUpload() method that is used to specify the image to be published.

Now, supposing that the image is in the webcam-shot.jpg file, we can do our post with the following command:

```
root@beaglebone:~# ./post_pic.php
post_pic.php[ 62]: name is "Open Graph Test User" (119617175051891)
post_pic.php[ 78]: done
```

To verify the post, we can log in with the test user account generated previously by selecting the **Log in as this test user** entry in the **Edit** button. The following screenshot shows what I got on my test:

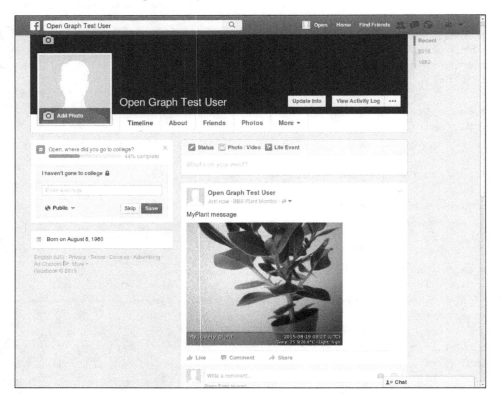

When finished, you have to log out from this test account and then log in again with your own account.

Now we are ready to go. Click on the **Status & Review** menu in the **Do you want to make this app and all its live features available to the general public?** section and click on **Yes**. Then, we need an access token for our Facebook profile, so we have to go to the **Tools & Support** top menu entry and then select the **Graph API Explorer** entry. On the new page, click on the **Graph API Explorer** drop-down list and then select our new application, **BBB Plant Monitor**. Then, click on the **Get Token** drop-down list and select the **Get Access Token** entry.

In the new window, you can add whatever permission you wish; however, we need only the extended permission publish_actions, so enable it and then click on the **Get Access Token** button.

At this time, an authorization window such as the one shown in the following screenshot should appear (sorry for the non-English text):

Then, approve and select the audience of the new app (just you, your friends only, or public. Don't worry about it—this setting can be changed later from your profile settings).

Now, get the new access token and copy it into the `config.php` file and then rerun the `get_info.php` script. If everything works well, you should get something like the following:

```
root@beaglebone:~# ./get_info.php
get_info.php[  62]: name is "Rodolfo Giometti" (10206138948545992)
get_info.php[  67]: permissions are:
get_info.php[  69]:     publish_actions is granted
get_info.php[  69]:     public_profile is granted
```

Great, now I can try to post a picture to my Facebook timeline by using the `post_pic.php` script, as follows:

```
root@beaglebone:~# ./post_pic.php
post_pic.php[  62]: name is "Rodolfo Giometti" (10206138948545992)
post_pic.php[  78]: done
```

The following screenshot shows a snippet of my Facebook timeline with the new post:

Okay, it seems to work well! However, there is a problem. If you click on the **i** in the blue circle (near the beginning of your new access token), a window will appear showing the token's expiration time (that is, the time till your access token will work), and, as you can see, it's very short! Usually, it is only 1 or 2 hours.

See the following screenshot for an example:

To increase this time, we can click on the **Open in Access Token Tool** button and then ask for an extended version by clicking on the **Extend Access Token** button. Then, a new extended token will be released, as shown in the following screenshot (note that it may happen that the system will request your profile password again before giving the extended token to you; if so, enter the password again):

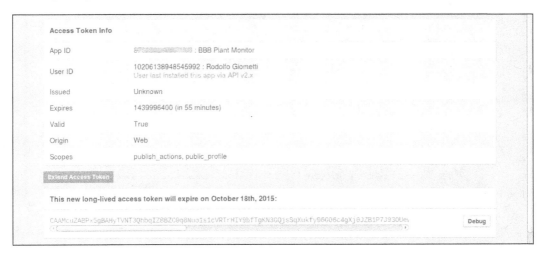

The new token is now valid for 60 days.

Unluckily, I've not found a suitable way to either automate this procedure or to renew a token when the current one expires. When it happens, you have to repeat this procedure to get a new valid token and then update the `config.php` file accordingly.

It may happen that you get an error in executing the command. In that case, you can enable the debugging messages by using the `-d` option argument and then trying to execute the command again, as follows:

```
root@beaglebone:~# ./get_info.php -d
get_info.php[ 54]: error!
get_info.php[ 55]: =====================================
=====================================
get_info.php[ 56]: exception 'Facebook\Exceptions\
FacebookSDKException' with message 'SSL certificate
problem: certificate is not yet valid' in /root/
chapter_07/facebook-php-sdk-v4-5.0-dev/src/Facebook/
HttpClients/FacebookCurlHttpClient.php:83
Stack trace:
#0 /root/chapter_07/facebook-php-sdk-v4-5.0-dev/src/
Facebook/FacebookClient.php(216): Facebook\HttpClients\
FacebookCurlHttpClient->send('https://graph.f...',
'GET', '', Array, 60)
#1 /root/chapter_07/facebook-php-sdk-v4-5.0-dev/src/
Facebook/Facebook.php(504): Facebook\FacebookClient-
>sendRequest(Object(Facebook\FacebookRequest))
#2 /root/chapter_07/facebook-php-sdk-v4-5.0-dev/
src/Facebook/Facebook.php(377): Facebook\Facebook-
>sendRequest('GET', '/me', Array, NULL, NULL, NULL)
#3 /root/chapter_07/get_info.php(52): Facebook\Facebook-
>get('/me')
#4 {main}get_info.php[ 57]: =============================
=================================================
```

This is not related to the Facebook API itself, but to an authentication stage of the SSL certificate. A possible workaround is implemented by using the following patch. However, this decreases the security level of the whole API. You are warned!

```
root@beaglebone# diff -u facebook-php-sdk-v4-5.0-dev/
src/Facebook/HttpClients/FacebookCurlHttpClient.php.orig
facebook-php-sdk-v4-5.0-dev/src/Facebook/HttpClients/
FacebookCurlHttpClient.php
--- facebook-php-sdk-v4-5.0-dev/src/Facebook/
HttpClients/FacebookCurlHttpClient.php.orig    2014-04-26
01:34:08.187500961 +0000
+++ facebook-php-sdk-v4-5.0-dev/src/Facebook/
HttpClients/FacebookCurlHttpClient.php    2014-04-26
01:34:37.582032215 +0000
@@ -111,7 +111,7 @@
            CURLOPT_RETURNTRANSFER => true, // Follow
301 redirects
            CURLOPT_HEADER => true, // Enable header
processing
            CURLOPT_SSL_VERIFYHOST => 2,
-           CURLOPT_SSL_VERIFYPEER => true,
+           CURLOPT_SSL_VERIFYPEER => false,
            CURLOPT_CAINFO => __DIR__ . '/certs/
DigiCertHighAssuranceEVRootCA.pem',
        ];
```

The monitoring loop

Now that the Facebook API is up and running, we can start writing the code to implement our plant monitor.

As already stated, the monitoring system is so slow that using a Bash script can be the most *quick-and-dirty* way to resolve our problem. In fact, we simply need to read plant's data from all installed sensors and then do some simple actions. The most difficult part is to create a status file containing all measured data to be exchanged with the web interface (see the next section for this last part).

The monitoring loop is in the `chapter_07/plant_mon.sh` file in the book's example code repository, while the relevant code is shown in the following snippet:

```
function daemon_body () {
    # Read plant data and take the first picture
    read_sensors
    next_date=$(do_picture)

    # The main loop
    dbg "start main loop"
    while sleep 1 ; do
        # Read plant data from all sensors
        read_sensors

        ( # Wait for lock on LOCK_FILE (fd 99) for 10 seconds
        flock -w 10 -x 99 || exit 1

        # Read the user parameters
        cff_mois=$(cat $STATUS_FILE | json_decode cff_mois)
        [ -z "$cff_mois" ] && cff_mois=1
        dbg "cff_mois=$cff_mois"

        # Compute the moisture level
        est_mois=$msr_mois
        if (( $(bc <<< "$int_temp < $ext_temp") == 1 )) ; then
            est_mois=$(bc -l <<< "$msr_mois + $cff_mois * ( $ext_temp -
$int_temp )")
        fi
        dbg "est_mois=$est_mois"

        # Write back the plant parameters
        json_encode lig_levl $lig_levl \
            int_temp $int_temp \
            ext_temp $ext_temp \
            msr_mois $msr_mois \
            cff_mois $cff_mois \
            est_mois $est_mois > $STATUS_FILE

        # Release the lock
        ) 99>$LOCK_FILE

        # Have to take a new picture?
        [ $(date "+%H%M") == "$next_date" ] && next_date=$(do_
picture)
    done
}
```

As you can see, there are three main steps to execute:

1. Read the plant's data from all sensors.

2. Exchange the data by managing the file addressed by the STATUS_FILE variable.

3. Take a new picture according to the user input.

The first step is implemented by the read_sensors function as follows:

```
function read_sensors ( ) {
    lig_levl=$(adc_read $LIGHT_DEV)
    int_temp=$(w1_read $INT_TEMP_DEV)
    ext_temp=$(i2c_read $EXT_TEMP_DEV)
    msr_mois=$(adc_read $MOISTURE_DEV)
    dbg "lig_levl=$lig_levl int_temp=$int_temp ext_temp=$ext_temp
msr_mois=$msr_mois"
    dbg "curr_date=$(date "+%H%M") next_date=$next_date"
}
```

The last step is implemented by the do_picture function as follows:

```
function do_picture ( ) {
    # Compute the light level
    ll="mid"
    [ $lig_levl -le $LIGHT_LOW ] && ll="low"
    [ $lig_levl -ge $LIGHT_HIGH ] && ll="high"

    # Take the picture
    fswebcam -q --title "My lovely plant" \
        --info "Temp: $ext_temp/$int_temp°C - Light: $ll" \
        --jpeg 85 $IMG_FILE

    # Compute the next picture time
    date -d "now +$1 minutes" "+%H%M"
}
```

The second action needs some explanation. As already stated, we have to send the plant's data to the web interface. From the web interface, we have to read an input value (the moisture coefficient). To do that, I decided to use a normal file due to the fact that all operations are executed very slowly.

The only difficulty when using a file to exchange data is related to the fact that we must be sure to have exclusive access to it. To do so, we can use the flock() system call to ask the system for exclusive access to a file in order to exclude concurrent reads or writes.

To use the `flock()` in a Bash script, we have the `flock` command that, when used in a shell script, must be used as suggested by the `flock`'s man pages. The code is as follows:

```
(
    flock -n 9 || exit 1
    # ... commands executed under lock ...
) 9>/var/lock/mylockfile
```

 See the `man flock` command for further information.

After grabbing the lock, we can start reading the file. It holds the system status in the JSON format because the file content must be shared with a PHP application that has predefined functions to manage such format. So, first we read the user input by using the `json_decode` function, then we calculate the estimated moisture level using the formula described at the beginning of this chapter, and, as the last step, we write back the status file in the JSON format by using the `json_encode` function.

To execute the program enabling all debugging messages, we can use the following command line:

```
root@beaglebone:~# ./plant_mon.sh -d -l -f -k 1
plant_mon.sh: min=1
plant_mon.sh: signals traps installed
plant_mon.sh: lig_levl=295 int_temp=29.50 ext_temp=31.01 msr_mois=0
plant_mon.sh: curr_date=1109 next_date=
plant_mon.sh: start main loop
plant_mon.sh: lig_levl=304 int_temp=29.50 ext_temp=31.11 msr_mois=0
plant_mon.sh: curr_date=1109 next_date=1110
plant_mon.sh: cff_mois=50
plant_mon.sh: est_mois=80.50
...
```

Then, the program can be stopped by running the *CTRL + C* keys.

 Remember to modify the line where I define the `INT_TEMP_DEV` variable holding the 1-Wire temperature sensor's ID according to the ID of your sensor, otherwise you're going to get a read error while executing the program.

Note that as the first step, the program reads all the sensors' data and then takes a picture in such a way that the web interface has whatever it needs to display the current system status to the user, as explained in the next section.

The web interface

This time, I'm going to implement a simple web interface by using the HTML, PHP, and JavaScript languages. What I want to implement is something really simple that allows the user to see the plant's data and the last picture of the plant stored on the system. Then, they should be able to post that picture on the Facebook timeline.

The core of the web interface is in the `chapter_07/plant.html` file in the book's example code repository, while the relevant code is shown in the following snippet:

```
<body>
 <h1>Plant monitor status</h1>

 <h2>Internal variables</h2>

 <table class="status">
  <tr class="d0">
    <td>Light level</td>
    <td><b id="lig_levl">0</b></td>
    <td></td>
  </tr>
  <tr class="d1">
    <td>Internal temperature[C]</td>
    <td><b id="int_temp">0</b></td>
    <td></td>
  </tr>
  <tr class="d0">
    <td>External temperature[C]</td>
    <td><b id="ext_temp">0</b></td>
    <td></td>
  </tr>
  <tr class="d1">
    <td>Measured moisture</td>
    <td><b id="msr_mois">0</b></td>
    <td></td>
  </tr>
  <tr class="d0">
    <td>Moisture coefficient</td>
    <td><b id="cff_mois">0</b></td>
```

```
    <td><input id="val_cff_mois" name="val_cff_mois" class="set-
inputbox">
      <button id="set_cff_mois" class="set-button">Set</button>
      </td>
  </tr>
  <tr class="d1">
    <td>Estimated moisture</td>
    <td><b id="est_mois">0</b></td>
    <td></td>
  </tr>
</table>

<h2>Last picture</h2>

<img id="webcam_shot" alt="Last picture">

<p><button id="post_pic" class="do-button">Post on Facebook</
button></p>

</body>
```

As you can see, there is a simple table where all the data is reported and where the user can change the moisture coefficient with the **Set** button. Then, the plant's picture is shown at the bottom with a **Post on Facebook** button to allow the user to post the current picture on Facebook.

The **Set** button is managed by the following JavaScript code included in the `plant.html` file:

```
<script>
 $(function() {
  $('button[class="set-button"]').click(function() {
    var id = $(this).attr("id");
    var box = document.getElementById(id.replace('set_', 'val_'));

    $.ajax({
     url: "/handler.php",
     type: "POST",
     data: "set=" + id.replace('set_', '') + "&val=" + box.value,
success: function() {
        console.log('set POST success');
      },
      error: function() {
       console.log('set POST error');
      }
```

```
    });
   });
  });
</script>
```

So, each time we press the button, we generate a POST request holding the user input value.

In a similar manner, the **Post on Facebook** button is managed by the following code:

```
<script>
  $(function() {
    $('button[class="do-button"]').click(function() {
      var id = $(this).attr("id");

      $.ajax({
        url: "/handler.php",
        type: "POST",
        data: "do=" + id,
        success: function() {
          console.log('do POST success');
        },
        error: function() {
          console.log('do POST error');
        }
      });
    });
  });
</script>
```

In this case, we generate another POST request, but with different parameters.

On the other hand, the following JavaScript code is used for two main tasks:

```
<script>
  var polldata = function() {
    $.getJSON('/handler.php', function(data) {
      $.each(data, function(key, val) {
        var e = document.getElementById(key);

        if (e != null) {
          if (e.type == "text")
            e.value = val;
          else
            e.textContent = val;
```

```
            }
        });
    });

    var url = '/webcam-shot.jpg';
    var d = new Date();
    $('#webcam_shot').attr('src', url + '?d=' + d.getTime());
    };

    setInterval(polldata, 1000);
</script>
```

The first task is to request and then update the plant's data on the web page, while the second task is to update the plant's image.

Note that we use a trick to force the browser to refresh the plant's image:

```
var url = '/webcam-shot.jpg';
var d = new Date();
$('#webcam_shot').attr('src', url + '?d=' +
d.getTime());
```

Here, I appended a parameter with the current date to the image's attribute in order to force the browser to update the image.

Looking at the preceding codes, the reader can notice that when the PHP file is executed, the handler.php script is executed in turn too. The handler.php file manages the data on the server side, and the following snippet shows its relevant code:

```
#
# Ok, do the job
#

# Check the POST requests
if (isset($_POST["val"]))
    $new_cff_mois = floatval($_POST["val"]);
else if (isset($_POST["do"]))
    do_post();

# Wait for lock on /tmp/plant.lock
$lock = file_lock(LOCK_FILE);
if (!$lock)
    die();
```

```
# Read the status file and decode it
$ret = file_get_contents(STATUS_FILE);
if ($ret === false)
   die();
$data = json_decode($ret, true);

# Use the stored value reset to a specific defualt
if (isset($new_cff_mois))
   $data['cff_mois'] = $new_cff_mois;

# Write back the new status (if needed)
$status = json_encode($data);
if (isset($new_cff_mois)) {
   $ret = file_put_contents(STATUS_FILE, $status);
   if ($ret === false)
      die();
}

# Release the lock
file_unlock($lock);

# Encode data for JSON
echo $status;
```

 The complete code can be found in the `chapter_07/handler.php` file in the book's example code repository.

At the first step, we check for any POST requests and, in case, we serve them. In the first case, we update the moisture coefficient, while in the second case we call the do_post() function that is explained next to post the plant's image on Facebook.

Then, we have to read (and eventually update) the system's status file by using the flock() system call. In PHP, the file lock is managed by the flock() function, as follows, in order to acquire and release the lock on a file:

```
function file_lock($name)
{
   $f = fopen($name, 'w');
   if ($f === false)
      return false;

   $ret = flock($f, LOCK_EX);
   if ($ret === false)
```

```
        return false;

    return $f;
}

function file_unlock($f)
{
    flock($f, LOCK_UN);
    fclose($f);
}
```

As the last action, we return the plant's data to the browser in the JSON format suitable for the calling JavaScript.

Okay, now we're at the last thing to show, which is the do_post() function:

```
function do_post()
{
    # Define the Facebook session
    $fb = new Facebook\Facebook([
        'app_id'                => APP_ID,
        'app_secret'            => APP_SECRET,
        'default_graph_version' => 'v2.4',
        'default_access_token'  => DEF_TOKEN,
        'fileUpload'            => true,
        'cookie'                => true,
    ]);

    # Publish to user's timeline
    $ret = $fb->post('/me/photos', array(
        'message'    => 'My lovely plant!',
        'source'     => $fb->videoToUpload(realpath('webcam-shot.jpg')),
    ));
}
```

This function simply performs the same steps as the post_pic.php script shown previously in order to post the plant's picture on the user's Facebook timeline.

Final test

To test the prototype, I first executed the chapter_07/SYSINIT.sh file in the book's example code repository to set up all peripherals:

root@beaglebone:~# ./SYSINIT.sh

done!

Now, after checking that the web server is running, I started the `plant_mon.sh` plant monitor script, enabling all debugging messages:

```
root@beaglebone:~# root@beaglebone:~/chapter_07# ./plant_mon.sh -d -1 -f
plant_mon.sh: min=10
plant_mon.sh: signals traps installed
plant_mon.sh: lig_levl=442 int_temp=29.50 ext_temp=29.91 msr_mois=0
plant_mon.sh: curr_date=0010 next_date=
plant_mon.sh: start main loop
plant_mon.sh: lig_levl=428 int_temp=29.50 ext_temp=29.25 msr_mois=0
plant_mon.sh: curr_date=0010 next_date=0011
plant_mon.sh: cff_mois=50
plant_mon.sh: est_mois=221
plant_mon.sh: lig_levl=423 int_temp=29.50 ext_temp=27.99 msr_mois=0
plant_mon.sh: curr_date=0010 next_date=0011
plant_mon.sh: cff_mois=50
plant_mon.sh: est_mois=220
...
```

Then, I set up the web server's root directory in order to implement the web interface. On my BeagleBone Black, the web server's root directory is `/var/www/`, but it may vary according to your system settings.

 The reader can take a look at the book *BeagleBone Essentials*, *Packt Publishing*, written by the author of this book, in order to get more information regarding how to set up a web server on the BeagleBone Black.

If you have the same configuration as mine, and the `plant_mon.sh` script is running, then your `/var/www/` directory should look like the following:

```
root@beaglebone:~# ls /var/www/
plant.lock  plant.status  webcam-shot.jpg
```

These files are created by the monitoring script, and they're, respectively, the lock file, the system's status file, and the last picture taken. Along with these files you also need the Facebook API (so we have to unzip its source code here, as shown previously) and the configuration files `config.php` and `setup.php`.

Then, we have to add the `plant.html`, `plant.css`, and `handler.php` files for the the web interface with the `jquery-1.9.1.js` file that can be downloaded from `https://code.jquery.com/jquery/` by using the browser on the host PC or directly on your BeagleBone Black with the following command line:

```
root@beaglebone:# wget --no-check-certificate https://code.jquery.com/
jquery-1.9.1.js
```

Then, we must make sure that all files are owned by the system user `www-user` in order to allow the web server to read/write them without problems. To do so, we can use the following command:

```
root@beaglebone:# cd /var/www && chown -R www-data:www-data *
```

If everything works well, your web server's root directory should look like the following:

```
root@beaglebone:/var/www# ls -l
total 308
-rw-r--r-- 1 www-data www-data    344 Aug 19  2015 config.php
drwxr-xr-x 5 www-data www-data   4096 Aug 18  2015 facebook-php-sdk-v4-
5.0-dev
-rw-r--r-- 1 www-data www-data   1846 Aug 26  2015 handler.php
-rw-r--r-- 1 www-data www-data 268381 Oct 24  2014 jquery-1.9.1.js
-rw-r--r-- 1 www-data www-data   2968 Aug 26  2015 plant.html
-rw-rw-rw- 1 www-data www-data      0 Apr 26 01:17 plant.lock
-rw-rw-rw- 1 www-data www-data     95 Apr 26 01:17 plant.status
-rw-r--r-- 1 www-data www-data    183 Aug 24  2015 setup.php
-rw-r--r-- 1 www-data www-data  17583 Apr 26 01:17 webcam-shot.jpg
```

Now, everything should be in place, so, on my host PC, I pointed my browser to the BeagleBone Black's IP address on the emulated Ethernet line via a USB cable to display the web interface. A screenshot of my test is as follows:

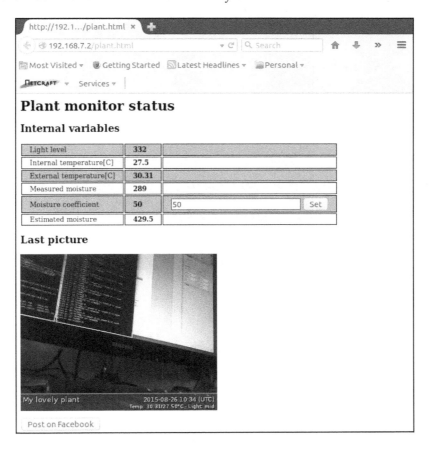

Notice that a similar result can be obtained by using a smartphone or tablet. In fact, if I connect my BeagleBone Black to my LAN and then point my smartphone's browser to the BeagleBone Black IP address, I get what is shown in the following screenshot:

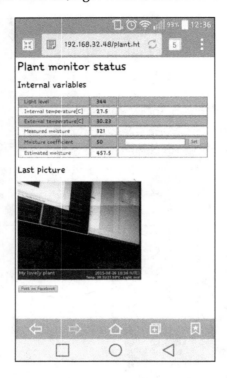

Remember that the IP address of the USB emulated Ethernet is usually `192.168.7.2`, while the IP address that the BeagleBone Black gets when it is connected to your LAN may vary according to your LAN settings. It can be retrieved by using the `ifconfig eth0` command on a BeagleBone Black's terminal.

Summary

This time, we used several sensors to get important data about our lovely plant. Then, we also discovered a simple way to exchange this data between processes by using a normal file. We learned how to use the Facebook PHP API to do a post on a user's timeline by using a simple script.

In the next chapter, we'll try to implement an intrusion alarm system with motion detection sensors that, in case of alarm, will start taking pictures of the intruders and then send them to the user's e-mail address.

Intrusion Detection System

8

Nowadays intrusion detection systems are quite common but really expensive. In this chapter, I'm going to show how we can implement a cheap intrusion detection system with a reasonable quality using our BeagleBone Black and two (or more) webcams.

The system will be able to alert the user by sending an e-mail with a photo of the intruder.

The basics of functioning

As mentioned earlier, we're going to use two webcams connected with our BeagleBone Black via a USB bus. Then, we'll install and run a special motion detection software that is able to detect a movement into a dynamic scene. When the program detects a movement, it will take one or more photos of the moving object and then send the pictures via e-mail to a user's account.

Setting up the hardware

This time, the connections are very simple since they are just done using several USB cables.

In the previous chapters, we have seen how to set up a webcam (see *Chapter 3, Aquarium Monitor*, for instance); but this time, we have a different configuration due to the fact that we're using two webcams at the same time.

As the reader might know, the BeagleBone Black board has only one USB host port, so to connect two webcams, we need a USB hub. These devices (used to connect more than one device to a USB host port) are very common, and the reader can find them anywhere on the Internet.

In theory, the more ports the hub has, the more webcams we can use in our system! But, of course, there is a maximum limit of usable webcams due to the fact that each webcam adds a CPU load to the system.

A little schematic of the system using a **USB HUB** with three ports is shown in the following diagram:

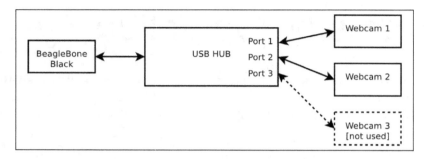

Setting up the webcams

For my prototype, I used two generic webcams supported by the **Video4Linux** driver class, as already explained in *Chapter 3*, *Aquarium Monitor*, connected to the **USB HUB** as shown in the following image. However, you can use your preferred device since it's a very common device.

 The curious reader can get more information about the USB hub's drivers at https://en.wikipedia.org/wiki/USB_hub.

To verify if everything is well connected and supported, you have to connect the webcams as shown in the diagram in the previous section. Then, you should get something similar to the output of my system, as follows:

```
root@beaglebone:~# ls -l /dev/video*
crw-rw---T 1 root video 81, 0 Jan  1  2000 /dev/video0
crw-rw---T 1 root video 81, 1 Jan  1  2000 /dev/video1
```

Note that you must use an external power supply for the HUB, or else your BeagleBone Black will be unable to supply enough current to manage both webcams.

Okay, now we can verify if the webcams are correctly managed using the `fswebcam` program in the same manner as in the previous chapter. However, this time we must specify which webcam must be used to take a simple picture to the `fswebcam` program. The trick can be done using the `-d` option argument, as shown in the following command line:

```
root@beaglebone:~# fswebcam -d /dev/video0 video0-shot.jpg
--- Opening /dev/video0...
Trying source module v4l2...
/dev/video0 opened.
No input was specified, using the first.
Adjusting resolution from 384x288 to 352x288.
--- Capturing frame...
Captured frame in 0.00 seconds.
--- Processing captured image...
Writing JPEG image to 'video0-shot.jpg'.
```

As already stated in *Chapter 3, Aquarium Monitor*, if you get a completely blank image with a message as follows, you can resolve the issue by adding the `-S` option argument to the command line:

```
root@beaglebone:~# fswebcam -d /dev/video0 -S 10 webcam-
shot.jpg
```

Then, to take a picture from the other webcam, we can use the following command line:

```
root@beaglebone:~# fswebcam -d /dev/video1 video1-shot.jpg
--- Opening /dev/video1...
Trying source module v4l2...
/dev/video1 opened.
No input was specified, using the first.
Adjusting resolution from 384x288 to 320x240.
--- Capturing frame...
Captured frame in 0.00 seconds.
--- Processing captured image...
Writing JPEG image to 'video1-shot.jpg'.
```

The following two screenshots show the two pictures from the two webcams, which are set face to face, that is, the first webcam gets a picture of the second one and vice versa.

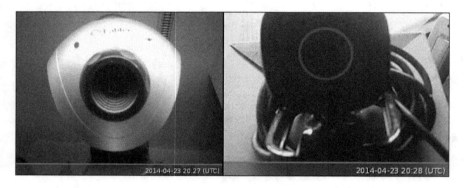

The final picture

The following screenshot shows all the devices that I used to implement this project and to test the software:

Nothing special to highlight here; all the connections are just very simple USB connections. However, let me highlight the fact that I used an external power supply for both the BeagleBone Black board and for the USB hub in order to avoid power loss due to the high power consumption from the webcams.

Setting up the software

This time, we have to set up two programs: the mailer and the motion detection system. The former, used to send the alarm e-mail message, is very simple and easy to set up; while the latter, used to implement the intrusions detection system, is a bit more complicated due to the fact that it supports tons of different devices and features.

Setting up the mailer

As requested by this project, we should alert the user about possible intrusions by sending them an e-mail. There exists several ways to send an e-mail on a UNIX-like system, and the most commonly used is the `mail` command that is called with the following command:

```
echo "Test message" | mail -s "test mail" email_address@somedomain.com
```

For further information about the mail program, the reader can take a look at its man pages (using the man mail command) or start reading about it at https://en.wikipedia.org/wiki/Mail_%28Unix%29.

The real problem is that this command relays on the system mailer, which is the real program that actually sends our letters over the Internet! By default, our BeagleBone Black has no valid mailer, so, if we try to send an e-mail message with the mail program, we get the following error:

```
root@beaglebone:~# echo "Test message" | mail -s "test mail"giometti@hce-
engineering.com
root@beaglebone:~# /usr/lib/sendmail: No such file or directory
"/root/dead.letter" 8/200
. . . message not sent.
```

 The -s option argument used in the preceding output is to specify a subject for the e-mail.

To solve our problem, we have to install a valid /usr/lib/sendmail program, and, as already stated, there are several ways to do it. I decided to use the ssmtp tool with my Gmail account.

 Note that the ssmtp tool is a generic tool to be used with a mailhost, so it is not a Gmail-related product. For further information on the tool, you can take a look at https://wiki.debian.org/sSMTP.

To install it, we can use the usual aptitude command with two other useful tools for e-mail processing:

```
root@beaglebone:~# aptitude install ssmtp mailutils mpack
```

After the installation, we have to modify the /etc/ssmtp/ssmtp.conf configuration file according to the following patch:

```
--- /etc/ssmtp/ssmtp.conf.orig   2015-09-11 21:52:39.531475392 +0000
+++ /etc/ssmtp/ssmtp.conf   2015-09-11 21:56:01.859600416 +0000
@@ -18,4 +18,10 @@
 # Are users allowed to set their own From: address?
```

```
 # YES - Allow the user to specify their own From: address
 # NO - Use the system generated From: address
-#FromLineOverride=YES
+FromLineOverride=YES
+
+# Add GMail settings
+mailhub=smtp.gmail.com:587
+AuthUser=rodolfo.giometti@gmail.com
+AuthPass=XXXXXXXXX
+useSTARTTLS=YES
```

The `FromLineOverride` setting has been enabled since we wish to specify our own `From:` address; then, the other fields are needed so we can send an e-mail message via our Gmail account.

> For obvious reasons, I replaced my password with the XXXXXXXXX string. Of course, you have to set up the `AuthUser` and `AuthPass` settings to suite your Gmail account.

If everything works well, we should now be able to send an e-mail using the following command:

```
root@beaglebone:~# echo "Test message" | mail -s "Test subject" -r "BBB
Guardian <myaccount@gmail.com>" giometti@hce-engineering.com
```

> Note that if your Gmail credentials are not correctly set up, you may get the following error message:
>
> ```
> send-mail: Authorization failed (535 5.7.8
> https://support.google.com/mail/answer/14257
> fr10sm1091535wib.14 - gsmtp)
> ```
>
> Also, note also that the `-r` option argument is used to specify a sender name; so, in the preceding example, in the `From:` field is displayed the `BBB Guardian <myaccount@gmail.com>` string, otherwise, the Gmail address is displayed instead.

The following screenshot shows the message as received on my smartphone:

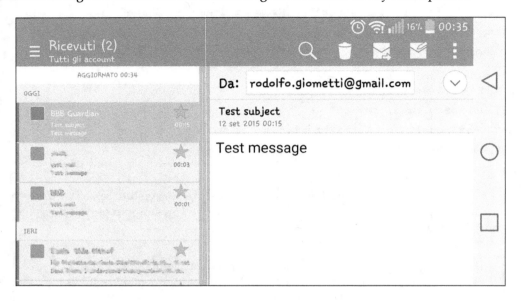

Using motion

> *Motion is a program that monitors the video signal from cameras. It is able to detect if a significant part of the picture has changed; in other words, it can detect motion.*
>
> *[Motion WebHome]*

 Visit the project homepage is at `http://www.lavrsen.dk/foswiki/bin/view/Motion/WebHome`.

The software is a libre CCTV software application developed for GNU/Linux-based systems, and as stated on the program's home site, it can monitor the video signal from one or more cameras and is able to detect if a significant part of the picture has changed, saving video when it detects that motion is occurring.

The program is a command-line-driven tool written in **C** and made for the Video4Linux interface. It can run as a daemon with a rather small footprint and low CPU usage. It can call to user configurable *triggers* when certain events occur, and then it generates either pictures (`.jpeg`, `.netpbm`) or videos (`.mpeg`, `.avi`).

`motion` is operated mainly via configuration files though the end video streams that can be viewed from a web browser.

Downloading the code

Downloading and installing `motion` on the BeagleBone Black is quite simple, since we simply have to use the usual command to install a new package as follows:

```
root@beaglebone:~# aptitude install motion
...
Setting up motion (3.2.12-3.4) ...
Adding group `motion' (GID 117) ...
Done.
Adding system user `motion' (UID 111) ...
Adding new user `motion' (UID 111) with group `motion' ...
Not creating home directory `/home/motion'.
Adding user `motion' to group `video' ...
Adding user motion to group video
Done.
[ ok ] Starting motion (via systemctl): motion.service.
```

When all the code is installed, it's time to configure the program! In fact, if we take a look at the system's log messages, we see the following output:

```
root@beaglebone:~# tail -f /var/log/syslog
...
Sep  4 15:18:41 beaglebone motion[4511]: Not starting motion daemon,
disabled via /etc/default/motion ... (warning).
```

The daemon is disabled by default due to the fact that it must be correctly configured before enabling it. So, let's see how we can do this in the next section.

Configuring the daemon

To configure the daemon in order to use two webcams, we have to modify three files: the main configuration file `/etc/motion/motion.conf`, the configuration files of each webcam `/etc/motion/thread1.conf`, and `/etc/motion/thread2.conf`. The daemon creates one thread per webcams used, and all special settings referring to a webcam must be set inside the corresponding file.

Let's start by modifying the `/etc/motion/motion.conf` file. First of all, we must enable one thread per webcam, so we have to apply the following patch:

```
--- motion.conf.orig   2014-04-23 21:12:18.511719124 +0000
+++ motion.conf    2014-04-23 21:12:47.710937877 +0000
@@ -630,8 +630,8 @@
```

```
# This motion.conf file AND thread1.conf and thread2.conf.
# Only put the options that are unique to each camera in the
# thread config files.
-; thread /usr/local/etc/thread1.conf
-; thread /usr/local/etc/thread2.conf
+thread /etc/motion/thread1.conf
+thread /etc/motion/thread2.conf
 ; thread /usr/local/etc/thread3.conf
 ; thread /usr/local/etc/thread4.conf
```

Then, we can verify the setting by running the `motion` daemon in debugging mode with the following command:

```
root@beaglebone:~# motion -s -n
[0] Processing thread 0 - config file /etc/motion/motion.conf
[0] Processing config file /etc/motion/thread1.conf
[0] Processing config file /etc/motion/thread2.conf
[0] Motion 3.2.12 Started
[0] ffmpeg LIBAVCODEC_BUILD 3482368 LIBAVFORMAT_BUILD 3478785
[0] Motion running in setup mode.
[0] Thread 1 is from /etc/motion/thread1.conf
[0] Thread 1 is device: /dev/video0 input 8
[0] Webcam port 8081
[0] Thread 2 is from /etc/motion/thread2.conf
[0] Thread 2 is device: /dev/video1 input 1
[0] Webcam port 8082
[0] Waiting for threads to finish, pid: 3096
[1] Thread 1 started
[0] motion-httpd/3.2.12 running, accepting connections
[0] motion-httpd: waiting for data on port TCP 8080
[2] Thread 2 started
[1] cap.driver: "uvcvideo"
[1] cap.card: "Microsoft LifeCam VX-800"
[1] cap.bus_info: "usb-musb-hdrc.1.auto-1.1"
[1] cap.capabilities=0x84000001
[1] - VIDEO_CAPTURE
[1] - STREAMING
[1] Config palette index 8 (YU12) doesn't work.
[1] Supported palettes:
```

```
[1] 0: YUYV (YUV 4:2:2 (YUYV))
[1] Selected palette YUYV
[1] Test palette YUYV (320x240)
[1] Using palette YUYV (320x240) bytesperlines 640 sizeimage 153600
colorspace 00000000
[1] found control 0x00980900, "Brightness", range -10,10
[1]    "Brightness", default 2, current 2
[1] found control 0x00980901, "Contrast", range 0,20
[1]    "Contrast", default 10, current 10
[1] found control 0x00980902, "Saturation", range 0,10
[1]    "Saturation", default 4, current 4
[1] found control 0x00980903, "Hue", range -5,5
[1]    "Hue", default 0, current 0
[1] found control 0x00980910, "Gamma", range 100,200
[1]    "Gamma", default 130, current 130
[1] found control 0x00980913, "Gain", range 32,48
[1]    "Gain", default 34, current 34
[1] mmap information:
[1] frames=4
[1] 0 length=153600
[1] 1 length=153600
[1] 2 length=153600
[1] 3 length=153600
[1] Using V4L2
[2] cap.driver: "gspca_zc3xx"
[2] cap.card: "USB Camera (046d:08a2)"
[2] cap.bus_info: "usb-musb-hdrc.1.auto-1.2"
[2] cap.capabilities=0x85000001
[2] - VIDEO_CAPTURE
[2] - READWRITE
[2] - STREAMING
[2] Unable to query input 1 VIDIOC_ENUMINPUT: Invalid argument
[2] ioctl (VIDIOCGCAP): Inappropriate ioctl for device
[2] Could not fetch initial image from camera
[2] Motion continues using width and height from config file(s)
[1] Resizing pre_capture buffer to 1 items
[2] Resizing pre_capture buffer to 1 items
[2] Started stream webcam server in port 8082
```

As we can see from the preceding output, we can get a lot of useful information about the daemon status. First of all, we notice that each line begins with a number in square brackets that address per thread output. The number 0 is for the `motion` main thread, the number 1 is for the first thread connected to the first webcam (device `/dev/video0`), and the number 2 is for the second thread connected to the second webcam (device `/dev/video1`).

Then, we see that for the first webcam, the daemon says gives us the following output:

```
[1] cap.driver: "uvcvideo"
[1] cap.card: "Microsoft LifeCam VX-800"
[1] cap.bus_info: "usb-musb-hdrc.1.auto-1.1"
[1] cap.capabilities=0x84000001
[1] - VIDEO_CAPTURE
[1] - STREAMING
[1] Config palette index 8 (YU12) doesn't work.
[1] Supported palettes:
[1] 0: YUYV (YUV 4:2:2 (YUYV))
[1] Selected palette YUYV
```

That is the current palette setting (YU12) is not valid for the webcam and the system says that it is going to use YUYV.

And an error message is displayed for thread 2:

```
[2] cap.driver: "gspca_zc3xx"
[2] cap.card: "USB Camera (046d:08a2)"
[2] cap.bus_info: "usb-musb-hdrc.1.auto-1.2"
[2] cap.capabilities=0x85000001
[2] - VIDEO_CAPTURE
[2] - READWRITE
[2] - STREAMING
[2] Unable to query input 1 VIDIOC_ENUMINPUT: Invalid argument
[2] ioctl (VIDIOCGCAP): Inappropriate ioctl for device
[2] Could not fetch initial image from camera
```

This time, it seems like a severe error, but let's go step by step and fix the first camera. In the `/etc/motion/thread1.conf` file, we see the following settings (the following is just a snippet of the whole file):

```
# Videodevice to be used for capturing  (default /dev/video0)
# for FreeBSD default is /dev/bktr0
```

```
videodevice /dev/video0

# The video input to be used (default: 8)
# Should normally be set to 1 for video/TV cards, and 8 for USB
cameras
input 8
```

The `videodevice` and `input` settings are correct, but the video palette setting is missing, so the default one is used. As seen in the preceding output, it's wrong. To fix it, we must add the following lines:

```
--- /etc/motion/thread1.conf.orig   2014-04-23 21:12:25.712890999 +0000
+++ /etc/motion/thread1.conf   2014-04-23 20:25:15.089843787 +0000
@@ -12,6 +12,25 @@
 # for FreeBSD default is /dev/bktr0
 videodevice /dev/video0

+# v4l2_palette allows to choose preferable palette to be use by motion
+# to capture from those supported by your videodevice. (default: 8)
+# E.g. if your videodevice supports both V4L2_PIX_FMT_SBGGR8 and
+# V4L2_PIX_FMT_MJPEG then motion will by default use V4L2_PIX_FMT_MJPEG.
+# Setting v4l2_palette to 1 forces motion to use V4L2_PIX_FMT_SBGGR8
+# instead.
+#
+# Values :
+# V4L2_PIX_FMT_SN9C10X : 0    'S910'
+# V4L2_PIX_FMT_SBGGR8  : 1    'BA81'
+# V4L2_PIX_FMT_MJPEG   : 2    'MJPEG'
+# V4L2_PIX_FMT_JPEG    : 3    'JPEG'
+# V4L2_PIX_FMT_RGB24   : 4    'RGB3'
+# V4L2_PIX_FMT_UYVY    : 5    'UYVY'
+# V4L2_PIX_FMT_YUYV    : 6    'YUYV'
+# V4L2_PIX_FMT_YUV422P : 7    '422P'
+# V4L2_PIX_FMT_YUV420  : 8    'YU12'
+v4l2_palette 8
+
 # The video input to be used (default: 8)
 # Should normally be set to 1 for video/TV cards, and 8 for USB cameras
 input 8
```

Note that I set the entry `v412_palette` to `6` in order to select the YUYV palette. Now, if we rerun the daemon, we get the following output:

```
[2]  Thread 2 started
[1]  cap.driver: "uvcvideo"
[1]  cap.card: "Microsoft LifeCam VX-800"
[1]  cap.bus_info: "usb-musb-hdrc.1.auto-1.1"
[1]  cap.capabilities=0x84000001
[1]  - VIDEO_CAPTURE
[1]  - STREAMING
[1]  Test palette YUYV (320x240)
[1]  Using palette YUYV (320x240) bytesperlines 640 sizeimage 153600
colorspace 00000000
```

Great! Now, let's fix the configuration file for the second webcam. In the `/etc/motion/thread2.conf` file, we see the following output:

```
# Videodevice to be used for capturing   (default /dev/video0)
# for FreeBSD default is /dev/bktr0
videodevice /dev/video1

# The video input to be used (default: 8)
# Should normally be set to 1 for video/TV cards, and 8 for USB
cameras
input 1
```

Again, the `videodevice` setting is correct, but the `input` setting is not! So, let's fix it as shown in the following patch and then rerun the daemon:

```
--- /etc/motion/thread2.conf.orig   2014-04-23 21:12:30.703125375 +0000
+++ /etc/motion/thread2.conf    2014-04-23 20:31:54.214843835 +0000
@@ -14,7 +14,7 @@

 # The video input to be used (default: 8)
 # Should normally be set to 1 for video/TV cards, and 8 for USB cameras
-input 1
+input 6

 # Draw a user defined text on the images using same options as C
#function strftime(3)
 # Default: Not defined = no text
```

Now, the daemon output for the second thread is changed as follows:

```
[2] cap.driver: "gspca_zc3xx"
[2] cap.card: "USB Camera (046d:08a2)"
[2] cap.bus_info: "usb-musb-hdrc.1.auto-1.2"
[2] cap.capabilities=0x85000001
[2] - VIDEO_CAPTURE
[2] - READWRITE
[2] - STREAMING
[2] Config palette index 8 (YU12) doesn't work.
[2] Supported palettes:
[1] Resizing pre_capture buffer to 1 items
[2] 0: JPEG (JPEG)
[2] Selected palette JPEG
```

So, we have to modify the /etc/motion/thread2.conf file again as shown in the following patch:

```
--- /etc/motion/thread2.conf.orig   2014-04-23 20:34:51.173828231 +0000
+++ /etc/motion/thread2.conf   2014-04-23 20:34:32.744140729 +0000
@@ -12,6 +12,25 @@
 # for FreeBSD default is /dev/bktr0
 videodevice /dev/video1


+# v4l2_palette allows to choose preferable palette to be use by motion
+# to capture from those supported by your videodevice. (default: 8)
+# E.g. if your videodevice supports both V4L2_PIX_FMT_SBGGR8 and
+# V4L2_PIX_FMT_MJPEG then motion will by default use V4L2_PIX_FMT_MJPEG.
+# Setting v4l2_palette to 1 forces motion to use V4L2_PIX_FMT_SBGGR8
+# instead.
+#
+# Values :
+# V4L2_PIX_FMT_SN9C10X : 0   'S910'
+# V4L2_PIX_FMT_SBGGR8  : 1   'BA81'
+# V4L2_PIX_FMT_MJPEG   : 2   'MJPEG'
+# V4L2_PIX_FMT_JPEG    : 3   'JPEG'
+# V4L2_PIX_FMT_RGB24   : 4   'RGB3'
```

```
+# V4L2_PIX_FMT_UYVY    : 5   'UYVY'
+# V4L2_PIX_FMT_YUYV    : 6   'YUYV'
+# V4L2_PIX_FMT_YUV422P : 7   '422P'
+# V4L2_PIX_FMT_YUV420  : 8   'YU12'
+v4l2_palette 3
+
 # The video input to be used (default: 8)
 # Should normally be set to 1 for video/TV cards, and 8 for USB cameras
 input 8
```

Now, if we rerun the daemon for the second thread, we get the following output:

```
[2] cap.driver: "gspca_zc3xx"
[2] cap.card: "USB Camera (046d:08a2)"
[2] cap.bus_info: "usb-musb-hdrc.1.auto-1.2"
[2] cap.capabilities=0x85000001
[2] - VIDEO_CAPTURE
[2] - READWRITE
[2] - STREAMING
[2] Test palette JPEG (320x240)
[2] Using palette JPEG (320x240) bytesperlines 320 sizeimage 29390
colorspace 00000007
```

Perfect! The webcams are now correctly configured.

The web interface

Now it's time to verify the video output by directly seeing a video stream. To do it, `motion` sets up several web servers to be used to monitor the main thread (thread numbered 0) and the per camera threads (threads numbered from 1 to N).

If we take a look at the `webcam_port` settings in the `motion.conf`, `thread1.conf` and `thread2.conf` files, we see that each thread opens a different monitoring port, as follows:

```
root@beaglebone:~# grep webcam_port /etc/motion/{motion,thread1,thread2}.
conf
/etc/motion/motion.conf:webcam_port 8081
/etc/motion/thread1.conf:webcam_port 8081
/etc/motion/thread2.conf:webcam_port 8082
```

The only settings that must be modified are `control_localhost` and `webcam_localhost`, which must be set to `off` in order to allow a remote control connection for the first thread and a remote webcam connection for the threads of each webcam. The patch is as follows:

```
--- /etc/motion/motion.conf.orig   2014-04-23 21:12:18.511719124 +0000
+++ /etc/motion/motion.conf        2014-04-23 20:48:18.068359577 +0000
@@ -410,7 +410,7 @@
 webcam_maxrate 1


 # Restrict webcam connections to localhost only (default: on)
-webcam_localhost on
+webcam_localhost off


 # Limits the number of images per connection (default: 0 = unlimited)
 # Number can be defined by multiplying actual webcam rate by desired
number of seconds
@@ -426,7 +426,7 @@
 control_port 8080


 # Restrict control connections to localhost only (default: on)
-control_localhost on
+control_localhost off


 # Output for http server, select off to choose raw text plain (default:
on)
 control_html_output on
```

Note that the daemon doesn't start if the `8080` port is occupied by another running process (such as Apache, for instance). Verify that this is not this case.

Now, if we rerun the daemon, we can verify that the three `motion` web servers are running at the `8080`, `8081`, and `8082` ports using the following command line in a different terminal:

```
root@beaglebone:~# netstat -pnl | grep motion
tcp        0      0 0.0.0.0:8080          0.0.0.0:*
LISTEN     2388/motion
tcp        0      0 0.0.0.0:8081          0.0.0.0:*
LISTEN     2388/motion
tcp        0      0 0.0.0.0:8082          0.0.0.0:*
LISTEN     2388/motion
```

Great! Now, we can use a normal browser to connect to the main thread (thread number 0), but for us to check the webcams' output, we can get the per webcam video stream at `http://192.168.7.2:8081` and `http://192.168.7.2:8082`, as shown in the following screenshots:

Note that in this last test, I executed the daemon without the `-s` option argument in order to disable the setup mode, that is, using the following command line:

`root@beaglebone:~# motion -n`

This is because I noticed that in the setup mode, the webcams work with a very bad video output (I don't know if this is a bug or a feature).

On the other hand, the control thread can be controlled via a web browser at `http://192.168.7.2:8080`. The following screenshot shows the main page:

If we navigate to the **All** | **Config** | **list** menu entries, we reach
`http://192.168.7.2:8080/0/config/list`, where we can get a page with all the
configuration settings of the main thread, as shown in the following screenshot:

Note that we are able to change each setting just by clicking on the
relevant link and entering the new value. However, we're not going
to use these interfaces to set up the system in this book.

As for the main thread, we can get the configuration of each running thread by just clicking on the relative link and then navigating to the menu. As an example, for the thread 1, we can read its current configuration at `http://192.168.7.2:8080/1/config/list`, as shown in the following screenshot:

Managing events

Now it's time to see how we can perform some actions when an event occurs. The `motion` daemon defines several events, all reported in the main configuration file. In fact, in the `/etc/motion/motion.conf` file, we see the following settings (again a snippet of the file):

```
###########################################################
# External Commands, Warnings and Logging:
# You can use conversion specifiers for the on_xxxx commands
# %Y = year, %m = month, %d = date,
# %H = hour, %M = minute, %S = second,
# %v = event, %q = frame number, %t = thread (camera) number,
# %D = changed pixels, %N = noise level,
# %i and %J = width and height of motion area,
# %K and %L = X and Y coordinates of motion center
# %C = value defined by text_event
# %f = filename with full path
# %n = number indicating filetype
# Both %f and %n are only defined for on_picture_save,
```

```
# on_movie_start and on_movie_end
# Quotation marks round string are allowed.
##########################################################

# Do not sound beeps when detecting motion (default: on)
# Note: motion never beeps when running in daemon mode.
quiet on

# Command to be executed when an event starts. (default: none)
# An event starts at first motion detected after a period of no motion
defined by gap
; on_event_start value

# Command to be executed when an event ends after a period of no
motion
# (default: none). The period of no motion is defined by option gap.
; on_event_end value

# Command to be executed when a picture (.ppm|.jpg) is saved (default:
none)
# To give the filename as an argument to a command append it with %f
; on_picture_save value

# Command to be executed when a motion frame is detected (default:
none)
; on_motion_detected value

# Command to be executed when motion in a predefined area is detected
# Check option 'area_detect'.   (default: none)
; on_area_detected value

# Command to be executed when a movie file (.mpg|.avi) is created.
(default: none)
# To give the filename as an argument to a command append it with %f
; on_movie_start value

# Command to be executed when a movie file (.mpg|.avi) is closed.
(default: none)
# To give the filename as an argument to a command append it with %f
; on_movie_end value

# Command to be executed when a camera can't be opened or if it is
lost
```

```
# NOTE: There is situations when motion doesn't detect a lost camera!
# It depends on the driver, some drivers don't detect a lost camera at
all
# Some hang the motion thread. Some even hang the PC! (default: none)
; on_camera_lost value
```

These are all the possible events reported by the daemon, and here we can define the command to execute when one event occurs. We just have to enter a command file with specific arguments, and the daemon will call it at the right time. The allowed arguments are shown in the comment at the top of the preceding list.

As an example, and in order to better understand how the mechanism works, let's consider the following simple **Bash** script named `args.sh`:

```bash
#!/bin/bash

NAME=$(basename $0)
ID=$RANDOM

function log ( ) {
    echo "$(date "+%s.%N"): $NAME-$ID: $1"
}

log "executing with $# args"

n=1
for arg ; do
    log "$n) $arg"
    n=$((n + 1))
done

log "done"

exit 0
```

 The code is stored in the `chapter_08/bin/args.sh` script in the book's example code repository.

If we execute it from the command line, it simply prints its arguments (with a timestamp prefix) as follows:

```
root@beaglebone:~/chapter_08# ./args.sh arg1 "arg 2" "..." 'arg-N'
1398299270.472083231: args.sh-12334: executing with 4 args
1398299270.498241897: args.sh-12334: 1) arg1
```

```
1398299270.523545772: args.sh-12334: 2) arg 2
1398299270.548859939: args.sh-12334: 3) ...
1398299270.574398731: args.sh-12334: 4) arg-N
1398299270.599793272: args.sh-12334: done
```

 Note that the program prints a random number after its name. This is because we'll need a unique name to distinguish between thread 1 and thread 2 (see the following section).

Now, if we copy this script in the `/usr/local/bin/` directory, we can call it, as follows:

```
root@beaglebone:~/chapter_08# /usr/local/bin/args.sh test command line
1398299445.322540043: args.sh-13425: executing with 3 args
1398299445.348607251: args.sh-13425: 1) test
1398299445.374537126: args.sh-13425: 2) command
1398299445.400261585: args.sh-13425: 3) line
1398299445.429571918: args.sh-13425: done
```

We can use this program with `motion` in order to discover which arguments are passed to an external program when an event occurs. As an example, we can consider the `on_picture_save` event. We can enable it with the following patch:

```
--- /etc/motion/motion.conf.orig    2014-04-23 21:12:18.511719124 +0000
+++ /etc/motion/motion.conf    2015-09-11 20:58:19.334749400 +0000
@@ -518,7 +518,7 @@

 # Command to be executed when a picture (.ppm|.jpg) is saved (default:
#none)
 # To give the filename as an argument to a command append it with %f
-; on_picture_save value
+on_picture_save /usr/local/bin/args.sh %C %t %f

 # Command to be executed when a motion frame is detected (default: none)
 ; on_motion_detected value
```

In this manner, we ask `motion` to execute the `args.sh` script when a new picture is saved, passing to it the event's timestamp, the number of the thread that generated the event, and the full path name of the picture file.

Before running the daemon again, we must make sure that, in the per thread configuration file, the same event has been disabled by commenting on the relative line, as shown, for example, for the first thread, in the following patch:

```
--- /etc/motion/thread1.conf.orig    2014-04-23 21:12:25.712890999 +0000
+++ /etc/motion/thread1.conf    2015-09-11 20:57:49.828890021 +0000
@@ -50,7 +69,7 @@


 # Command to be executed when a picture (.ppm|.jpg) is saved (default:
none)
 # The filename of the picture is appended as an argument for the
command.
-on_picture_save /usr/local/motion-extras/camparse1.pl
+; on_picture_save /usr/local/motion-extras/camparse1.pl


 # Command to be executed when a movie file (.mpg|.avi) is closed.
 #(default: none)
```

> If you forget to disable the event in the webcams' threads, you'll get several errors, such as the following ones:
>
> ```
> [2] File of type 1 saved to: /usr/local/apache2/htdocs/
> cam2/01-20150911205458-00.jpg
>  &: 1:  &: /usr/local/motion-extras/camparse2.pl: not
> found
> [1] File of type 1 saved to: /usr/local/apache2/htdocs/
> cam1/01-20150911205555-01.jpg
>  &: 1:  &: /usr/local/motion-extras/camparse1.pl: not
> found
> ```
>
> So, remember to disable this setting for all running threads!

Now, if we execute the daemon again and do some movement in front of a camera, we get the following messages:

```
[1] File of type 1 saved to: /usr/local/apache2/htdocs/cam1/01-
20150911210615-00.jpg
1442005575.375926790: args.sh-7523: executing with 3 args
1442005575.386901248: args.sh-7523: 1) 20150911210615
1442005575.398043498: args.sh-7523: 2) 1
1442005575.408981165: args.sh-7523: 3) /usr/local/apache2/htdocs/cam1/01-
20150911210615-00.jpg
1442005575.419728082: args.sh-7523: done
```

Great! Everything is working correctly. Now, it's very easy to finish the job. In fact, we have to simply replace the args.sh script with a script that sends an e-mail to us with a picture attached! A snippet of a possible implementation of such program is as follows:

```
#
# Local functions
#

function log ( ) {
    echo "[$cam] $1"
}

function send_alert {
    # Build the attachments list
    [ ! -e $ALERT_LIST ] && return
    for f in $(head -n $ALERT_LIMIT $ALERT_LIST) ; do
        list="-a $f $list"
    done

    # Send the letter
    echo -e ${ALERT_MESG/\%time/$time} | \
        mail -s "$ALERT_SUBJ" -r "$ALERT_FROM" $list "$ALERT_TO"
}

usage() {
    echo "usage [to add image]: $NAME <timestamp><cam #><filepath>"
>&2
    echo "usage [to send alert]: $NAME <timestamp><cam #>" >&2
    exit 1
}

#
# Main
#
# Check command line
[ $# -lt 2 ] && usage

( # Wait for lock on LOCK_FILE (fd 99) for 10 seconds
flock -w 10 -x 99 || exit 1

if [ $# -eq 3 ] ; then
    # Must add the picture to the list
    time=$1
```

```
        cam=$2
        path=$3

        log "got new picture $path at $time"
        echo "$path" >> $ALERT_LIST

    elif [ $# -eq 2 ] ; then
        # Send the mail alert
        time=$1
        cam=$2

        log "sending alert at $time"
        send_alert
        rm $ALERT_LIST

    else

        cam="?"
        log "invalid command!"

    fi

    # Release the lock
    ) 99>$LOCK_FILE

    exit 0
```

 The complete code is stored in the `chapter_08/bin/send_alert.sh` script in the book's example code repository.

To use it, we have to copy it in `/usr/local/bin/`, as we did before, for the `args.sh` program. Then, we must replace all the occurrences of `args.sh` in the `/etc/motion/motion.conf` configuration file with `send_alert.sh`. When this is done, just rerun the `motion` daemon, and when a motion is detected, we should get a logging message as follows:

[1] File of type 1 saved to: /usr/local/apache2/htdocs/cam1/01-20150915210814-01.jpg

[1] got new picture /usr/local/apache2/htdocs/cam1/01-20150915210814-01.jpg at 20150915210814

[1] File of type 1 saved to: /usr/local/apache2/htdocs/cam1/01-20150915210815-00.jpg

```
[1] got new picture /usr/local/apache2/htdocs/cam1/01-20150915210815-00.
jpg at 20150915210814
```

```
[1] File of type 1 saved to: /usr/local/apache2/htdocs/cam1/01-
20150915210815-01.jpg
```

```
[1] got new picture /usr/local/apache2/htdocs/cam1/01-20150915210815-01.
jpg at 20150915210814
```

Note that it's quite common that a lot of pictures will be taken, so without some specific anti-flooding technique, we can risk to send tons of e-mails! The trick here is quite easy — using the gap and on_event_end options, we can generate an *e-mail send* event once motion decides that the current event is finished. In fact, we can see the following by taking a look at the configuration filed:

```
# Gap is the seconds of no motion detection that triggers the end of
an event
# An event is defined as a series of motion images taken within a
short timeframe.
# Recommended value is 60 seconds (Default). The value 0 is allowed
and disables
# events causing all Motion to be written to one single mpeg file and
no pre_capture.
gap 60
```

We can imagine storing the filenames of the images in a list, and then, when the on_event_end event occurs, we can read back the names and send one e-mail with attachments.

To enable the on_event_end event, I used to following setting:

```
--- /etc/motion/motion.conf.orig   2014-04-23 21:12:18.511719124 +0000
+++ /etc/motion/motion.conf   2015-09-15 20:55:31.654352880 +0000
@@ -514,11 +514,11 @@

 # Command to be executed when an event ends after a period of no motion
 # (default: none). The period of no motion is defined by option gap.
-; on_event_end value
+on_event_end /usr/local/bin/send_alert.sh %C %t

 # Command to be executed when a picture (.ppm|.jpg) is saved (default:
none)
 # To give the filename as an argument to a command append it with %f
-; on_picture_save value
+on_picture_save /usr/local/bin/send_alert.sh %C %t %f
```

Here, the `send_alert.sh` script implements this solution. If we run it without arguments, it displays a short usage message as follows:

```
root@beaglebone:~# ./send_alert.sh
usage [to add image]: send_alert.sh <timestamp> <cam #> <filepath>
usage [to send alert]: send_alert.sh <timestamp> <cam #>
```

If executed with three arguments, it stores `filepath` in the file addressed by the `ALERT_LIST` variable, while when it is executed with two arguments, it rereads the file and sends an e-mail with a number of pictures (limited by the variable `ALERT_LIMIT`) as attachments.

To test if the program works correctly, we can try to execute it with just two arguments, and then verify that an e-mail message arrives to our account.

Final test

Now it's time to test our prototype. To do it, I decided to point the webcams toward my library shelf, where my precious stamp collection is kept. Then, I run the `motion` utility and just wait.

Note that this time, there is nothing special to do to configure the hardware.

After a while, I receive the following e-mail:

Then, looking at one picture, I discovered a very dangerous intruder.

Summary

In this chapter, there was very little hardware work to do, but on the other hand, we discovered how to use a really powerful tool named motion. This tool allows us to realize (quasi) professional, if minimal, anti-intrusion system. Also, you learned how to send simple e-mails with attached pictures to inform the user about an important event in the system.

In the next chapter, we'll discover how to use different identification devices (such as RFID readers and smart card readers) to implement an access control system.

9

Twitter Access Control System with Smart Card and RFID

Identifying people or objects with a computer may seem like an easy task, but, in reality, there is a lot of technology behind the devices used to accomplish this task.

In this chapter, we are going to use a smart card reader and two kinds of RFID readers (a **Low Frequency** or **LF** reader suitable for short ranges and an **Ultra High Frequency** or **UHF** reader suitable for long ranges) in order to show different possibilities to implement a minimal identifying system for access control.

Once the detection has been done, the system will send a message to our Twitter account informing our followers about what is happening (in a normal case, we may lock or unlock something, but I decided to do something different).

The basics of functioning

The smart cards and the smart card readers, for example, are complex devices that are used everywhere nowadays, from our credit cards to our smartphones. The term smart card implies a set of technologies, including integrated circuits, microprocessors, memories, antennas, and so on, in the same integrated circuit, to form a microchip that is the heart of a smart card. On the other hand, the smart card readers are complex devices that can communicate with the cards and save data on it or return data to a computer.

The reader can get more information regarding the smart card world by taking a look at https://en.wikipedia.org/wiki/Smart_card.

The evolution of smart cards are the **Radio-Frequency Identification** (**RFID**) devices that can be used to identify people or objects in a contactless form, starting from a few centimeters to several meters. The RFID readers and the corresponding tags (or transponders) are high technology radio devices that can exchange data to each other in order to accomplish identification tasks.

 The reader can get more information regarding the **RFID** world by taking a look at https://en.wikipedia.org/wiki/Radio-frequency_identification.

This time, all the complexity of this project is inside the hardware devices (the smart card reader and the RFID readers) and their corresponding software managers, so we only have to write the code to get access to them and retrieve the data held in the smart cards or the RFID tags only.

In order to show different techniques to manage an identification device, we're going to write three programs (one per device) in three different programming languages. However, the result of all programs will be the same: when a well-defined person (or object) has been recognized, a message will be sent to our Twitter account.

To accomplish this last task, we're going to use a dedicated (and interesting) tool that allows the user to manage a Twitter account from the command line.

Every identification system has its own characteristics; however:

- The solution with the smart card reader can be used, where the identification can be done by inserting a credit card (or something similar) somewhere for identification. This is not suitable for wireless applications. The smart card reader I'm going to use in my prototype is a USB device with a slot where the smart card must be inserted.

- The second solution, that is, the one with the RFID LF reader, can be used where the identification tasks need wireless, but where the distance between the object to identify and the reader is no more than a few centimeters. These kinds of devices are usually very simple, such as the one I'm going to use in my prototype. The RFID reader is connected with the host by a serial port, and it simply returns a string each time a tag is detected.

- The last solution is implemented with a RFID UHF reader, that is, with a device that can detect tags in a wireless mode, as the preceding RFID LF reader does, but at a distance of several meters long. These UHF devices can be more complex than the LF ones, such as the one I'm going to use for my prototype. The RFID reader still uses a serial connection to talk with the host, but it implements a more elaborate protocol to exchange data.

Setting up the hardware

As just said in the previous section, this time we have to connect a USB device and two serial devices. Regarding the USB device, the main problem is that it has a nonstandard USB connector, so we have to find a trick to solve the problem (see the following part); while for the serial devices, we have to find two free serial ports on the BeagleBone Black's expansion connectors.

Regarding this last issue, we should remember that the BeagleBone Black has six on-board serial ports that are disabled by default, except the /dev/ttyO0 device, which is coupled to the serial console. If we do log in into the system, we can easily verify it by using the following command:

```
root@BeagleBone:~# ls -l /dev/ttyO*
crw-rw---- 1 root tty 248, 0 Apr 23 20:20 /dev/ttyO0
```

To enable the other serial ports, we need to modify the kernel settings in order to enable the serial port that we wish to use. The choice of which ports to enable depends on the pins we'd like to use to connect our devices, and the following table may help us in choosing them:

Device	TxD	RxD	RTS	CTS	Name
/dev/ttyO1	P9.24	P9.26			UART1
/dev/ttyO2	P9.21	P9.22	P8.38	P8.37	UART2
/dev/ttyO4	P9.13	P9.11	P8.33	P8.35	UART4
/dev/ttyO5	P8.37	P8.38			UART5

All the devices are suitable for our scope, so I choose to use the two /dev/ttyO1 and /dev/ttyO2 devices, and, to activate them, we can use the following commands:

```
root@BeagleBone:~# echo BB-UART1 > /sys/devices/bone_capemgr.9/slots
root@BeagleBone:~# echo BB-UART2 > /sys/devices/bone_capemgr.9/slots
```

Now, two new serial ports are ready to be used, as shown by the following command:

```
root@beaglebone:~# ls -l /dev/ttyO*
crw-rw---- 1 root tty      248, 0 Apr 23 20:20 /dev/ttyO0
crw-rw---T 1 root dialout 248, 1 Apr 23 21:48 /dev/ttyO1
crw-rw---T 1 root dialout 248, 2 Apr 23 21:48 /dev/ttyO2
```

The reader can also take a look at the book *BeagleBone Essentials*, *Packt Publishing*, written by the author of this book, in order to have more information regarding how to manage the BeagleBone Black's serial ports needed to communicate with the sensors.

Setting up the smart card reader

The smart card reader I used in this prototype is shown in the following image:

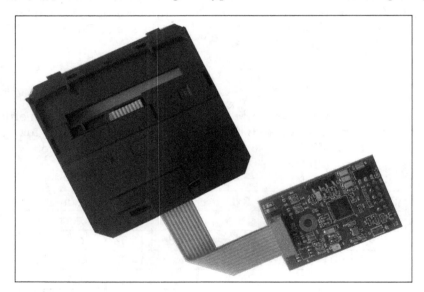

The device can be purchased at the following link (or by surfing the Internet): http://www.cosino.io/product/http://www.cosino.io/product/smartcard-reader-isoiec-7816.

The device is based on the chip **Maxim 73S1215F**, and its datasheet is available at http://datasheets.maximintegrated.com/en/ds/73S1215F.pdf.

As stated before, this device has a nonstandard USB connector, so we have to find a trick to connect it to our BeagleBone Black.

The *quick and dirty* solution can be in using a USB plug type A adapter from an old USB device, which then must be soldered with the board, as shown in the following image:

The connections must be done according to the following table:

Smart Card reader pin	USB plug type A cable
VBus	Red
D-	White
D+	Yellow
GND	Green

 The connector pin out can be retrieved at https://en.wikipedia.org/wiki/USB in the **pin out** box on the left.

If the connections are correct, once you connect the device to the BeagleBone Black, you should get an output as follows:

```
usb usb1: usb wakeup-resume
usb usb1: usb auto-resume
hub 1-0:1.0: hub_resume
hub 1-0:1.0: port 1: status 0101 change 0001
hub 1-0:1.0: state 7 ports 1 chg 0002 evt 0000
hub 1-0:1.0: port 1, status 0101, change 0000, 12 Mb/s
usb 1-1: new full-speed USB device number 2 using musb-hdrc
usb 1-1: ep0 maxpacket = 16
usb 1-1: skipped 1 descriptor after interface
usb 1-1: skipped 1 descriptor after interface
usb 1-1: default language 0x0409
usb 1-1: udev 2, busnum 1, minor = 1
usb 1-1: New USB device found, idVendor=1862, idProduct=0001
usb 1-1: New USB device strings: Mfr=1, Product=2, SerialNumber=3
usb 1-1: Product: TSC12xxF CCID-DFU Version 2.10
usb 1-1: Manufacturer: Teridian Semiconductors
usb 1-1: SerialNumber: 123456789
usb 1-1: usb_probe_device
usb 1-1: configuration #1 chosen from 1 choice
usb 1-1: adding 1-1:1.0 (config #1, interface 0)
usb 1-1: adding 1-1:1.1 (config #1, interface 1)
hub 1-0:1.0: state 7 ports 1 chg 0000 evt 0002
hub 1-0:1.0: port 1 enable change, status 00000103
```

Okay, everything works correctly, but now we need some packages to manage our smart card reader. So, let's install them by using the following command:

```
root@beaglebone:~# aptitude install pcsc-tools pcscd libccid
```

Once finished, the `pcsc` tool is ready to work.

 The curious reader may take a look at the following URL for further information on this tool: http://ludovic.rousseau.free.fr/ softwares/pcsc-tools/.

Once installed, we can execute it with the following command:

```
root@beaglebone:~# pcsc_scan
PC/SC device scanner
V 1.4.20 (c) 2001-2011, Ludovic Rousseau <ludovic.rousseau@free.fr>
Compiled with PC/SC lite version: 1.8.3
Using reader plug'n play mechanism
Scanning present readers...
Waiting for the first reader...
```

> In case you get the following error instead of the preceding output, you can try to restart the daemon with the `/etc/init.d/pcscd restart` command and then execute the `pcsc_scan` tool again:
>
> `SCardEstablishContext: Service not available.`

Okay, the daemon started correctly, but it still didn't recognize our device. In this case, we have to patch the `/etc/libccid_Info.plist` configuration file, as shown in the following patch:

```
--- /etc/libccid_Info.plist.orig    2014-04-23 20:39:48.664062641
+0000
+++ /etc/libccid_Info.plist    2014-04-23 20:40:28.705078271 +0000
@@ -325,6 +325,7 @@
        <string>0x08C3</string>
        <string>0x08C3</string>
        <string>0x15E1</string>
+       <string>0x1862</string>
    </array>

    <key>ifdProductID</key>
@@ -550,6 +551,7 @@
        <string>0x0401</string>
        <string>0x0402</string>
        <string>0x2007</string>
+       <string>0x0001</string>
    </array>

    <key>ifdFriendlyName</key>
@@ -775,6 +777,7 @@
        <string>Precise Biometrics Precise 250 MC</string>
        <string>Precise Biometrics Precise 200 MC</string>
```

```
        <string>RSA RSA SecurID (R) Authenticator</string>
+               <string>TSC12xxF</string>
    </array>

    <key>Copyright</key>
```

After all the modifications are in place, we have to restart the daemon. Now the output should change as follows:

```
root@beaglebone:~# /etc/init.d/pcscd restart
[ ok ] Restarting pcscd (via systemctl): pcscd.service.
root@beaglebone:~# pcsc_scan
PC/SC device scanner
V 1.4.20 (c) 2001-2011, Ludovic Rousseau <ludovic.rousseau@free.fr>
Compiled with PC/SC lite version: 1.8.3
Using reader plug'n play mechanism
Scanning present readers...
0: TSC12xxF (123456789) 00 00
1: TSC12xxF (123456789) 00 01
2: TSC12xxF (123456789) 00 02
3: TSC12xxF (123456789) 00 03
4: TSC12xxF (123456789) 00 04


Wed Apr 23 20:40:56 2014
Reader 0: TSC12xxF (123456789) 00 00
  Card state: Card removed,
Reader 1: TSC12xxF (123456789) 00 01
  Card state: Card removed,
Reader 2: TSC12xxF (123456789) 00 02
  Card state: Card removed,
Reader 3: TSC12xxF (123456789) 00 03
  Card state: Card removed,
Reader 4: TSC12xxF (123456789) 00 04
  Card state: Card removed,
```

Great! Now we can verify that the reader is really working by inserting a card into the socket and by verifying that the tool should print something, as follows:

```
Wed Apr 23 20:52:22 2014
Reader 0: TSC12xxF (123456789) 00 00
  Card state: Card inserted,
  ATR: 3B BE 11 00 00 41 01 38 00 00 00 00 00 00 00 01 90 00

ATR: 3B BE 11 00 00 41 01 38 00 00 00 00 00 00 00 01 90 00
+ TS = 3B --> Direct Convention
+ T0 = BE, Y(1): 1011, K: 14 (historical bytes)
  TA(1) = 11 --> Fi=372, Di=1, 372 cycles/ETU
    10752 bits/s at 4 MHz, fMax for Fi = 5 MHz => 13440 bits/s
  TB(1) = 00 --> VPP is not electrically connected
  TD(1) = 00 --> Y(i+1) = 0000, Protocol T = 0
-----
+ Historical bytes: 41 01 38 00 00 00 00 00 00 00 00 01 90 00
  Category indicator byte: 41 (proprietary format)

Possibly identified card (using /usr/share/pcsc/smartcard_list.txt):
3B BE 11 00 00 41 01 38 00 00 00 00 00 00 00 01 90 00
  ACS (Advanced Card System) ACOS-1
```

The device is functioning; however, we need a dedicated program to manage the cards. So, let's install the `python-pyscard` package with the usual `aptitude` command and then consider the following code snippet:

```
#
# Smart Card Observer
#

class printobserver(CardObserver):
    def update(self, observable, (addedcards, removedcards)):
        for card in addedcards:
            logging.info("->] " + toHexString(card.atr))
        for card in removedcards:
            logging.info("<-] " + toHexString(card.atr))

#
# The daemon body
```

```
    #

def daemon_body():
    # The main loop
    logging.info("INFO waiting for card... (hit CTRL+C to stop)")

    try:
        cardmonitor = CardMonitor()
        cardobserver = printobserver()
        cardmonitor.addObserver(cardobserver)

        while True:
            sleep(1000000) # sleep forever

    except:
        cardmonitor.deleteObserver(cardobserver)
```

 The complete code is stored in the chapter_09/smart_card/smart_ card.py script in the book's example code repository.

The program defines a cardmonitor object and then adds its observer with the addObserver() method in order to be called when a card is inserted or removed.

If executed, the program gives an output as follows:

```
root@beaglebone:~/smart_card# ./smart_card.py
INFO:root:INFO waiting for card... (hit CTRL+C to stop)
INFO:root:->] 3B BE 11 00 00 41 01 38 00 00 00 00 00 00 00 00 01 90 00
INFO:root:<-] 3B BE 11 00 00 41 01 38 00 00 00 00 00 00 00 01 90 00
```

 If you get the following error while executing the command, you need the python-daemon package:

ImportError: No module named daemon

You an resolve the issue by using the following command:

root@beaglebone:~/smart_card# pip install python-daemon

Setting up the RFID LF reader

As an RFID LF reader, we can use the device shown in the following image that sends its data through a serial port at the TTL 3.3V level:

The device can be purchased at the following link (or by surfing the Internet): http://www.cosino.io/product/lf-rfid-low-voltage-reader.

The datasheet of this device is available at http://cdn.sparkfun.com/datasheets/Sensors/ID/ID-2LA,%20ID-12LA,%20ID-20LA2013-4-10.pdf.

It can be directly connected to our BeagleBone Black to the following pins of the expansion connector *P9*, corresponding to the already-enabled serial device /dev/tty01:

Pins	RFID LF reader pins - label
P9.4 - Vcc	8 - Vcc
P9.26 - RxD	6 - TX
P9.2 - GND	7 - GND

After all the pins have been connected, the tag's data will be available on the /dev/ttyO1 device. To quickly verify it, we can use the following commands:

root@BeagleBone:~# stty -F /dev/ttyO1 9600 raw

root@BeagleBone:~# cat /dev/ttyO1

Then, when approaching a tag to the reader, we should hear a *beep*, and the corresponding tag's ID should appear to the command line as follows. (The following cat command is repeated from before for better readability, so you don't need to retype it):

root@BeagleBone:~# cat /dev/ttyO1

.6F007F4E1E40

However, using the cat command is not the best thing to do for our purposes since its output is not a completely clean ASCII text (see the device's datasheet for further information about this issue); in fact, some *dirty* bytes are received before the tag ID. For example, the dot '.' before the tag ID is one of these bytes. So, we can imagine writing a dedicated tool to clean the received messages from the device in order to have a clean ASCII ID string. A code snippet of such a tool is as follows:

```
# Read the tags' IDs
cat $dev | while read id ; do
    # Remove the non printable characters and print the data
    echo -n $id | tr '\r' '\n' | tr -cd '[:alnum:]\n'
done
```

 The complete code is stored in the chapter_09/rfid_lf/rfid_lf.sh script in the book's example code repository.

The cat command reads the data from the device addressed by the dev variable, as in the preceding example; then, the output is passed through the tr command in order to remove nonprintable characters. The result is as follows:

root@beaglebone:~/rfid_lf# ./rfid_lf.sh /dev/ttyO1

6F007F48C199

 The curious reader can take a look at the tr's man pages for further information about its usage.

Setting up the RFID UHF reader

As an RFID UHF reader, we can use the following device that sends its data through a serial port at the TTL 3.3V level:

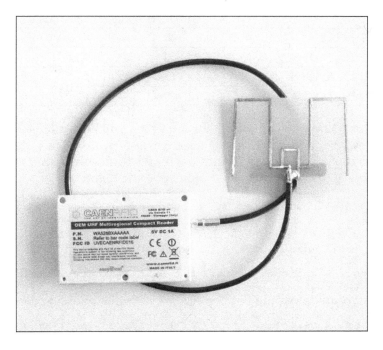

The device can be purchased at the following link (or by surfing the Internet): `http://www.cosino.io/product/uhf-rfid-long-range-reader`.

The product's information from the manufacturer is available at `http://www.caenrfid.it/en/CaenProd.jsp?mypage=3&parent=59&idmod=818`.

It can be directly connected to our BeagleBone Black to the following pins of the expansion connector *P9*, which are connected with the already-enabled serial device `/dev/ttyO2`:

Pins	RFID UHF reader pins - label
P9.6 - Vcc	1 - +5V
P9.21 - TxD	9 - RXD
P9.22 - RxD	10 - TXD
P9.1 - GND	12 - GND

After all the pins have been connected, the tag's data will be available at the `/dev/ttyO2` device, but to get them, we need extra software. In fact, this device requires a special protocol to communicate with the host, so we need to install a dedicated **C** library to do the trick, as explained in the following part.

We need to download, compile, and then install three libraries: `libmsgbuff`, `libavp`, and `libcaenrfid`.

First of all, we need some prerequisite packages. So, let's install them:

```
root@beaglebone:~# aptitude install git debhelper dctrl-tools
```

Now, we can start downloading the first library with the following command:

```
root@beaglebone:~# git clone http://github.com/cosino/libmsgbuff.git
```

Then, we have to enter in the new directory `libmsgbuff` and execute the `autogen.sh` command, as follows:

```
root@beaglebone:~# cd libmsgbuff
root@beaglebone:~/libmsgbuff# ./autogen.sh
```

It may happen that you get the following errors:

```
aclocal:configure.ac:11: warning: macro `AM_SILENT_RULES'
not found in library
aclocal:configure.ac:18: warning: macro `AM_PROG_AR' not
found in library
configure.ac:11: error: possibly undefined macro: AM_
SILENT_RULES
        If this token and others are legitimate, please use
m4_pattern_allow.
        See the Autoconf documentation.
```

In this case, the lines with the macros AM_SILENT_RULES and AM_PROG_AR should be deleted, as shown in the following patch:

```
    index dcfd1ce..333e417 100644
    --- a/configure.ac
    +++ b/configure.ac
    @@ -8,14 +8,12 @@ AC_CONFIG_SRCDIR([msgbuff.c])
    AC_CONFIG_HEADERS([configure.h])

    AM_INIT_AUTOMAKE([1.9 foreign -Wall -Werror])
    -AM_SILENT_RULES([yes])

    # Global settings
    AC_SUBST(EXTRA_CFLAGS, ['-Wall -D_GNU_SOURCE -include
    configure.h'])

    # Checks for programs
    AC_PROG_CXX
    -AM_PROG_AR
    AC_PROG_AWK
    AC_PROG_CC
    AC_PROG_CPP
```

Then, we can safely restart the autogen.sh command.

Then, to recompile the library, we can use the following command line:

```
root@beaglebone:~/libmsgbuff# ./debian/rules binary
dpkg-deb: building package `libmsgbuff0' in `../libmsgbuff0_0.60.0_armhf.
deb'.
dpkg-deb: building package `libmsgbuff-dev' in `../libmsgbuff-dev_0.60.0_
armhf.deb'.
```

Okay, now that the packages are ready we can install them by using the dpkg command, as follows:

```
root@beaglebone:~/libmsgbuff# dpkg -i ../libmsgbuff0_0.60.0_armhf.deb ../
libmsgbuff-dev_0.60.0_armhf.deb
Setting up libmsgbuff0 (0.60.0) ...
Setting up libmsgbuff-dev (0.60.0) ...
```

Now it's the turn of the second library. The steps are the same as in the preceding example. Once done, move to the parent directory and then download the new sources with the following git command:

```
root@beaglebone:~# git clone http://github.com/cosino/libavp.git
```

Then, execute the autogen.sh script in the library's directory:

```
root@beaglebone:~# cd libavp
root@beaglebone:~/libavp# ./autogen.sh
```

 Again, as in the preceding example, if an undefined macro error occurs, just apply the same patch as in the preceding example at the current autogen.sh script.

Then, start the following compilation:

```
root@beaglebone:~/libavp# ./debian/rules binary
dpkg-deb: building package `libavp0' in `../libavp0_0.80.0_armhf.deb'.
dpkg-deb: building package `libavp-dev' in `../libavp-dev_0.80.0_armhf.
deb'.
```

And, finally, execute the dpkg command to install the packages:

```
root@beaglebone:~/libavp# dpkg -i ../libavp0_0.80.0_armhf.deb ../libavp-
dev_0.80.0_armhf.deb
```

Okay, for the last library the procedure is similar, but with a little note. Move to the parent directory. Then, download the code and execute the autogen.sh script (patch it as before if needed):

```
root@beaglebone:~# git clone http://github.com/cosino/libcaenrfid.git
root@beaglebone:~# cd libcaenrfid/
root@beaglebone:~/libcaenrfid# ./autogen.sh
```

Then, we need to create two new files for the BeagleBone Black's architecture (which is named `armhf` in Debian). The commands are as follows:

```
root@beaglebone:~/libcaenrfid# cp src/linux-gnueabi.c src/linux-gnueabihf.c

root@beaglebone:~/libcaenrfid# cp src/linux-gnueabi.h src/linux-gnueabihf.h
```

Now we can execute the usual package generation command followed by the installation one, as follows:

```
root@beaglebone:~/libcaenrfid# ./debian/rules binary

...

dpkg-deb: building package `libcaenrfid0' in `../libcaenrfid0_0.91.0_armhf.deb'.

dpkg-deb: building package `libcaenrfid-dev' in `../libcaenrfid-dev_0.91.0_armhf.deb'.

root@beaglebone:~/libcaenrfid# dpkg -i ../libcaenrfid0_0.91.0_armhf.deb ../libcaenrfid-dev_0.91.0_armhf.deb
```

At this point, the needed libraries are in place, so we can compile our program to get access to the RFID UHF reader. A snippet of a possible implementation is as follows:

```c
int main(int argc, char *argv[])
{
    int i;
    struct caenrfid_handle handle;
    char string[] = "Source_0";
    struct caenrfid_tag *tag;
    size_t size;
    char *str;
    int ret;

    if (argc < 2)
        usage();

        /* Start a new connection with the CAENRFIDD server */
        ret = caenrfid_open(CAENRFID_PORT_RS232, argv[1], &handle);
        if (ret < 0)
            usage();

        /* Set session "S2" for logical source 0 */
        ret = caenrfid_set_srcconf(&handle, "Source_0",
            CAENRFID_SRC_CFG_G2_SESSION, 2);
```

```
        if (ret < 0) {
            err("cannot set session 2 (err=%d)", ret);
            exit(EXIT_FAILURE);
        }

        while (1) {
            /* Do the inventory */
            ret = caenrfid_inventory(&handle, string, &tag, &size);
            if (ret < 0) {
                err("cannot get data (err=%d)", ret);
                exit(EXIT_FAILURE);
            }

            /* Report results */
            for (i = 0; i < size; i++) {
                str = bin2hex(tag[i].id, tag[i].len);
                EXIT_ON(!str);

                info("%.*s %.*s %.*s %d",
                    tag[i].len * 2, str,
                    CAENRFID_SOURCE_NAME_LEN, tag[i].source,
                    CAENRFID_READPOINT_NAME_LEN, tag[i].readpoint,
                    tag[i].type);

                free(str);
            }

            /* Free inventory data */
            free(tag);
        }

        caenrfid_close(&handle);

        return 0;
    }
```

 The complete code is stored in the `chapter_09/rfid_uhf/`
`rfid_uhf.c` script in the book's example code repository.

The program simply uses the `caenrfid_open()` method to establish a connection
with the reader and the `caenrfid_inventory()` method to detect the tags. The
`caenrfid_set_srcconf()` method is used to set an internal special feature in order
to avoid multiple readings of the same tag.

The program can be compiled with the `make` command, executed in the `rfid_uhf` directory, and the tool can be used as follows:

```
root@beaglebone:~/rfid_uhf# ./rfid_uhf /dev/ttyO2
```

The program answers with no output in case there are no tags near the reader's antenna, but if we approach some tags we get something as follows:

```
root@beaglebone:~/chapter_09/rfid_uhf# ./rfid_uhf /dev/ttyO2
rfid_uhf.c[ 110]: main: e280113020002021dda500ab Source_0 Ant0 3
```

Note that, in this case, and contrary to what happens in the RFID LF case, the reader can detect a tag even a few meters away (the distance depends on the antenna you're using!)

The final picture

The following image shows the prototype I realized to implement this project and to test the software:

Note that to use the RFID UHF reader, you must use an external power supply, while for the other two readers it is not needed.

Setting up the software

After the hardware has been set up, most of the job is done; to finish our job, we need to first install a tool to get access to our Twitter account, and then we have to add a mechanism to call it each time a successful identification process is accomplished. So, in the following sections I'm going to show how to install and correctly set up a command line tool to communicate with Twitter and then how to call it in three different programming languages for three different identification systems.

To simplify the project a bit, we can use a static list of known IDs stored in each program, but you can understand that this list can be easily managed by an external database. So, I leave this implementation as an exercise for you.

Setting up the Twitter utility

The utility I'm going to use to get access to a Twitter account is named with the single character t. The t program, as reported on its home page, derives from the Twitter SMS commands:

> *The CLI takes syntactic cues from the Twitter SMS commands, but it offers vastly more commands and capabilities than are available via SMS.*

In fact, once installed, it uses simple commands to update our Twitter status, follow/unfollow users, retrieve detailed information about a Twitter user, create a list for everyone you're following, and so on.

 For a complete reference of the t tool, the `https://github.com/sferik/t` URL is a good starting point.

To install this tool into our BeagleBone Black, we first need the ruby-dev package with the aptitude program:

```
root@beaglebone:~# aptitude install ruby-dev
```

Then, t is installed with the following command:

```
root@beaglebone:~# gem install t -V
```

 The execution of this command can be very slow! So, be patient and wait.

Once the installation has ended, we can execute the program, and if everything works well, a long list of available commands should be displayed as follows:

```
root@beaglebone:~# t -h
Commands:
    t accounts                       # List accounts
    t authorize                      # Allows an application to request user...
    t block USER [USER...]           # Block users.
    t delete SUBCOMMAND ...ARGS      # Delete Tweets, Direct Messages, etc.
    t direct_messages                # Returns the 20 most recent Direct Mes...
    t direct_messages_sent           # Returns the 20 most recent Direct Mes...
    t dm USER MESSAGE                # Sends that person a Direct Message.
    t does_contain [USER/]LIST USER  # Find out whether a list contains a user.
    t does_follow USER [USER]        # Find out whether one user follows ano...

    ...
```

At this point, as done for other social networks, we have to create a special application for our Twitter account to get access to our data. To do so, let's point our browser to the `https://apps.twitter.com/app/new` URL. We'll see a form where we can fill out information about our new application. Simply fill in three fields: **Name**, **Description**, and **Website**. Note that the name of the application needs to be unique across all Twitter users and cannot contain the word `twitter`, while the website can be arbitrary (for instance, `http://www.mydomain.com`), as shown in the following screenshot:

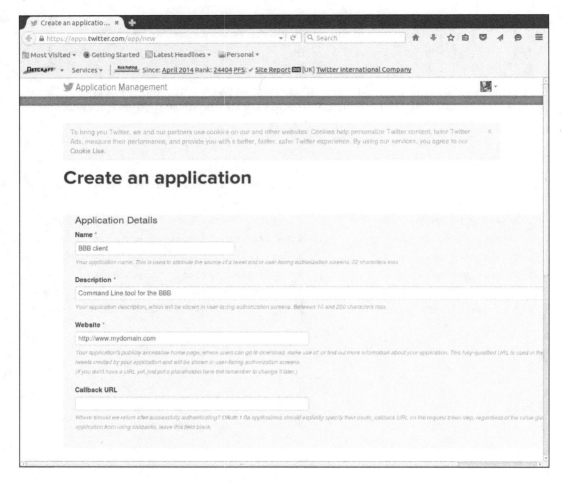

Regarding the **Callback URL** field, you can leave it blank. Then, click on the checkbox for developer terms agreement at the bottom of the page and then click on the **Create your Twitter application** button.

Once your application has been successfully created, you will see a page where you can manage your application settings, as shown in the following screenshot:

Now, go to the **Permissions** tab and change **Access** type to **Read, Write and Access direct messages** and save.

The next step is to authorize your application to access your Twitter account. For that, run the following command:

```
root@beaglebone:~# t authorize
Welcome! Before you can use t, you'll first need to register an
application with Twitter. Just follow the steps below:
  1. Sign in to the Twitter Application Management site and click
     "Create New App".
  2. Complete the required fields and submit the form.
     Note: Your application must have a unique name.
  3. Go to the Permissions tab of your application, and change the
     Access setting to "Read, Write and Access direct messages".
  4. Go to the API Keys tab to view the consumer key and secret,
     which you'll need to copy and paste below when prompted.


Press [Enter] to open the Twitter Developer site.
```

Then, once the *Enter* key has been pressed, the following output is shown:

```
xprop:  unable to open display ''
xprop:  unable to open display ''
Enter your API key: /usr/bin/xdg-open: 1: eval: www-browser: not found
/usr/bin/xdg-open: 1: eval: links2: not found
/usr/bin/xdg-open: 1: eval: elinks: not found
/usr/bin/xdg-open: 1: eval: links: not found
/usr/bin/xdg-open: 1: eval: lynx: not found
/usr/bin/xdg-open: 1: eval: w3m: not found
xdg-open: no method available for opening 'https://apps.twitter.com'
```

Apart from the error messages due to the fact that t cannot execute any browser at all, we have to go to the **Keys and Access Token** tab and enter the key in the **Consumer Key (API Key)** field located under **Application Settings**. Then, the tool will ask for the API secret, so you have to enter the **Consumer Secret (API Secret)** in the same page as before.

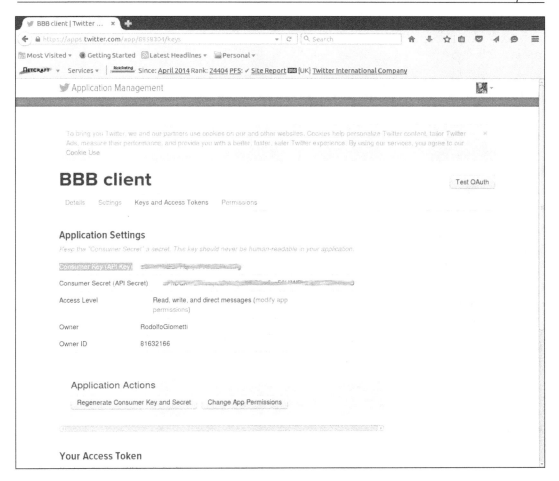

When finished, if both the keys are valid, the tool will display the following output:

```
In a moment, you will be directed to the Twitter app authorization page.
Perform the following steps to complete the authorization process:
  1. Sign in to Twitter.
  2. Press "Authorize app".
  3. Copy and paste the supplied PIN below when prompted.


Press [Enter] to open the Twitter app authorization page.
```

As before, the tool will try to open the browser again in order to show the Twitter application authorization page, but of course, it cannot, so the following error message is shown:

```
xprop:   unable to open display ''
xprop:   unable to open display ''
Enter the supplied PIN: /usr/bin/xdg-open: 1: eval: www-browser: not
found
/usr/bin/xdg-open: 1: eval: links2: not found
/usr/bin/xdg-open: 1: eval: elinks: not found
/usr/bin/xdg-open: 1: eval: links: not found
/usr/bin/xdg-open: 1: eval: lynx: not found
/usr/bin/xdg-open: 1: eval: w3m: not found
xdg-open: no method available for opening 'https://api.twitter.com/oauth/
authorize?oauth_callback=oob&oauth_consumer_key=sHSeFMEGPRqRyf9V0UB4LtQ
Og&oauth_nonce=9T9rSHXiaSiWXkh0ksVE5ioTcop0srz7xMG92VhVI&oauth_signatur
e=oNWj1Lj%225BUmrFkD%252B065axJv6WSeM%253D&oauth_signature_method=HMAC-
SHA1&oauth_timestamp=1443370645&oauth_token=J2fp-gAAAAAhyrAAABAUA-
YNw8&oauth_version=1.0'
```

Okay, we just need to *copy and paste* the preceding URL into our browser on the host PC to finish the job. To be clear, the URL is as follows:

```
https://api.twitter.com/oauth/authorize?oauth_callback=oob&oauth_
consumer_key=sHSeFMEGPRqRyf9V0UB4LtQOg&oauth_nonce=9T9rSHXiaSiWXkh0ksVE5
ioTcop0srz7xMG92VhVI&oauth_signature=oNWj1Lj%225BUmrFkD%252B065axJv6WSeM
%253D&oauth_signature_method=HMAC-SHA1&oauth_timestamp=1443370645&oauth_
token=J2fp-gAAAAAhyrAAABAUA-YNw8&oauth_version=1.0
```

Then, a new page where your Twitter credentials are to be put should appear, as shown in the following screenshot:

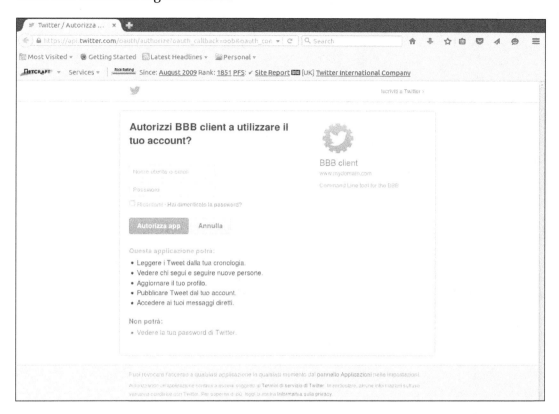

 Sorry for the Italian, but this is what my Twitter account's default language is set to.

Put your Twitter credentials and, if they are correct, the system should give you a PIN to be used to finish the authorization process (see the following screenshot):

Just *copy and paste* the PIN into the terminal where the tool is running and press *Enter* (again, you should not care about the error when launching the browser). However, if all steps are correct, the last message from t should be the following:

Authorization successful.

Great! Now, we are ready to do our first tweet from the BeagleBone Black's command line! The command is as follows:

```
root@beaglebone:~# t update 'Hello there! This is my first tweet from the
command line!'
Tweet posted by @RodolfoGiometti.

Run `t delete status 648174339569897474` to delete.
```

The following screenshot shows a snippet of my Twitter account where the recently sent message is published:

The smart card implementation

Let's now start with the first implementation of our identifying system by getting access to the smart card. The code is written in Python, and it shows a possible implementation of our access control system by using a smart card reader.

 Note that implementation is very minimal since we limit our attention to the ATR parameter, which cannot be used to uniquely identify a smart card in all circumstances.

The program is very similar to the one stored in the `chapter_09/smart_card/smart_card.py` file, so I'm going to show only the relevant differences here:

```
# The known IDs
ID2NAME = {
        '11 11 11 11 11 11 11 11 11 11 11 11 11 11 11 11 11 11 11':
"user1",
        '22 22 22 22 22 22 22 22 22 22 22 22 22 22 22 22 22 22 22':
"user2",
        '3B BE 11 00 00 41 01 38 00 00 00 00 00 00 00 01 90 00':
'Rodolfo Giometti'
}
...
#
# Smart Card Observer
#

class printobserver(CardObserver):
    def update(self, observable, (addedcards, removedcards)):
        for card in addedcards:
            try:
                id = toHexString(card.atr)
            except:
                pass
            if len(id) == 0:
                continue
            logging.info("got tag ID " + id)

            # Verify that the tag ID is known and then
            # tweet the event
```

```
try:
    name = ID2NAME[id]
except:
    logging.info("unknow tag ID! Ignored")
    continue

logging.info("Twitting that " + name + " was arrived!")
call([t_cmd, t_args, name + " was arrived!"])
```

 The complete code is stored in the `chapter_09/smart_card/smart_card2twitter.py` script in the book's example code repository.

The `ID2NAME` array holds a list of known IDs, that is, our *database* of valid IDs that are associated with well-known people. It's easy here to imagine that using a real database would be a better implementation, but this solution is fine for our teaching purposes.

The `update()` method extracts the smart card's ATR field, and then, instead of simply printing it, it compares the current ID with our internal database, and in case of positive match, it calls the `t` tool in order to update the Twitter account.

The RFID LF implementation

As in the preceding example, we have to modify the `chapter_09/rfid_lf/rfid_lf.sh` **Bash** script a bit in order to call the `t` tool if the current tag ID is found in the list of the known IDs held by the `ID2NAME` array. A snippet of the modified code is as follows:

```
# The known IDs
declare -gA 'ID2NAME=(
    [111111111111]="user1",
    [222222222222]="user2"
    [6F007F4E1E40]="Rodolfo Giometti"
)'
...
# Read the tags' IDs
cat $dev | while read id ; do
    # Remove the non printable characters
    id=$(echo $id | tr -cd '[:alnum:]')
    info "got tag ID $id"

    # Verify that the tag ID is known and then tweet the event
    name=${ID2NAME[$id]}
```

```
    if [ -z "$name" ] ; then
        info "unknow tag ID! Ignored"
    else
        info "Twitting that $name was arrived!"
        $t_cmd $t_args "$name was arrived!"
    fi
done
```

 The complete code is stored in the `chapter_09/rfid_lf/rfid_lf2twitter.sh` script in the book's example code repository.

The RFID UHF implementation

The last implementation is in C, and it uses the RFID UHF reader in order to take the identification string. The trick is now well-known; we simply need to modify the `chapter_09/rfid_uhf/rfid_uhf.c` program in order to check the current tag ID with the known ones held again in the well-known ID2NAME array. The code snippet is as follows:

```
/* The known IDs */
struct associative_array_s {
    char *id;
    char *name;
} ID2NAME[] = {
    { "111111111111111111111111", "user1" },
    { "222222222222222222222222", "user2" },
    { "e280113020002021dda500ab", "Rodolfo Giometti" },
};
...
    /* The main loop */
    while (1) {
        /* Do the inventory */
        ret = caenrfid_inventory(&handle, string, &tag, &size);
        if (ret < 0) {
            err("cannot get data (err=%d)", ret);
            exit(EXIT_FAILURE);
        }

        /* Report results */
        for (i = 0; i < size; i++) {
            str = bin2hex(tag[i].id, tag[i].len);
            EXIT_ON(!str);
```

```
            info("got tag ID %.*s", tag[i].len * 2, str);

            for (j = 0; j < ARRAY_SIZE(ID2NAME); j++)
                if (strncmp(str, ID2NAME[j].id,
                    tag[i].len * 2) == 0)
                break;
            if (j < ARRAY_SIZE(ID2NAME)) {
                info("Twitting that %s was arrived!",
                    ID2NAME[j].name);
                ret = asprintf(&cmd, "%s %s %s was arrived!",
                    t_cmd, t_arg, ID2NAME[j].name);
                EXIT_ON(ret < 1);
                ret = system(cmd);
                EXIT_ON(ret < 0);
                free(cmd);
            } else
                info("unknow tag ID! Ignored");

            free(str);
        }

        /* Free inventory data */
        free(tag);
    }
```

 The complete code is stored in the `chapter_09/rfid_uhf/rfid_uhf2twitter.c` file in the book's example code repository.

Before executing it, don't forget to compile it!

Final test

To test our prototype, we have to run the three different programs presented in the preceding section. As in the previous chapters, I first executed the `chapter_09/SYSINIT.sh` file in the book's example code repository as usual to set up all peripherals:

```
root@beaglebone:~# ./SYSINIT.sh
done!
```

Then, let's start the smart card program and insert the smartcard as follows:

```
root@beaglebone:~/smart_card# ./smart_card2twitter.py
INFO:root:got tag ID 3B BE 11 00 00 41 01 38 00 00 00 00 00 00 00 00
01 90 00
INFO:root:Twitting that Rodolfo Giometti was arrived!
Tweet posted by @RodolfoGiometti.

Run `t delete status 649586168313552896` to delete.
```

Now, press the *CTRL + C* keys to stop the program, and let's try the RFID LF program as follows by approaching the relative tag:

```
root@beaglebone:~/rfid_lf# ./rfid_lf2twitter.sh /dev/ttyO1
rfid_lf2twitter.sh: got tag ID 6F007F4E1E40
rfid_lf2twitter.sh: Twitting that Rodolfo Giometti was arrived!
Tweet posted by @RodolfoGiometti.

Run `t delete status 649586168313552896` to delete.
```

Again, stop the program with the *CTRL + C* keys, and let's test the last program as follows by approaching the relative tag again:

```
root@beaglebone:~/rfid_uhf# ./rfid_uhf2twitter /dev/ttyO2
rfid_uhf2twitter.c[ 122]: main: Twitting that Rodolfo Giometti was
arrived!
Tweet posted by @RodolfoGiometti.

Run `t delete status 649586168313552896` to delete.
```

Summary

In this chapter, we discovered how to publish a message to a Twitter account and three different ways to identify people or objects by using different identification technologies and programming languages.

In the next chapter, we'll discover how to manage some lights by using a common remote controller (or any infrared capable device). We'll see how our BeagleBone Black board can receive some commands by using our TV remote.

10

A Lights Manager with a TV Remote Controller

In this project, we will manage the lights of our home by using a normal TV remote controller.

In reality, we can use any remote controller we have, but the idea is to add remote controlling via the infrared mechanism to any device in our home. In fact, in this chapter I'm going to show how to manage a simple on/off device; but this concept can be easily extended to any other device we can connect to our BeagleBone Black!

We'll see how to capture the infrared messages that a remote controller sends to our BeagleBone Black by using a suitable circuitry, and then, we'll use a dedicated kernel driver in order to manage such messages and convert them into well-defined commands for our userspace programs.

The basics of functioning

The functioning of the prototype we're going to realize is quite simple. We need an electronic circuit that can detect the infrared light emitted by the remote controller and then generate some impulses that are caught by a special software that can detect and store them into a configuration file in order to be used at later time. Then, by using a special daemon, we can convert a button pressed on the remote controller into a suitable command for our BeagleBone Black.

In this scenario, the hardware we have to realize is very simple. We just need a small circuitry with an infrared-capable photo diode (infrared receiver). On the other hand, the software part is a bit more complicated due to the fact that we first need a kernel driver to reliably detect the message from the remote control, and then a user-level program to record it, a program to recognize which button has been pressed, and a last program to convert such pressures into on and off commands (or whatever we wish to control).

Due to lack of space, I'm going to manage the relays array used in *Chapter 3, Aquarium Monitor*, leaving it to you to connect whatever you want.

 Warning—remember that, even if the relays array used is suitable for controlling high voltages, *for safety reasons, you should not connect any device with voltages higher then 12V if you don't know what you are doing!*

Setting up the hardware

As stated before, the hardware setting is quite simple. The relays array has already been set up in a previous chapter, while the infrared receiver circuitry is really tricky. So, let's go ahead!

Setting up the infrared detector

The infrared detector (or receiver) I used in this prototype is shown in the following image (actually, the receiver is the device with the red dot; the other one is just a transmitter that we're not using here).

The devices can be purchased at the following link (or by surfing the Internet): http://www.cosino.io/product/infrared-emitter-detector.

The datasheet is available at https://www.sparkfun.com/datasheets/Components/LTR-301.pdf.

Note that the image shows the topmost part only of the infrared devices. In reality, they look similar to a normal diode.

The circuit to manage it is shown in the following diagram:

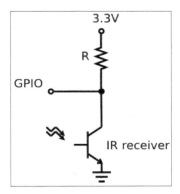

The **infrared receiver** (**IR**) is the diode with the red dot in the preceding diagram and **R** is a 6.8KΩ resistor. The following table shows the connections with the BeagleBone Black:

Pin	IR receiver label
P9.3 - Vcc	3.3V
P9.12 - GPIO60	GPIO @R
P9.1 - GND	GND

To test the functionality, we can set the GPIO 60 pin as an input pin by using the following command:

```
root@arm:~# ./bin/gpio_set.sh 60 in
```

Remember that you must not have any driver loaded into the system that uses that pin, otherwise you'll get an error message!

Then, we can use the following script to continuously read the GPIO status, and then print on the terminal when it goes to value 0:

```
root@arm:~# while true ; do cat /sys/class/gpio/gpio60/value | grep 0 ;
done
```

When you point your remote controller to the infrared receiver and press a button you, should get an output as follows:

```
root@arm:~# while true ; do cat /sys/class/gpio/gpio60/value | grep 0 ;
done
0
0
0
0
0
...
```

To stop the script, just hit the *CTRL + C* key sequence.

Setting up the relays array

The relays array is shown in the following image. It's the device already used in *Chapter 3*, *Aquarium Monitor*, so you can refer there for further information about it, while here I'm going to show the connections needed for this prototype only.

The device can be purchased at the following link (or by surfing the Internet): http://www.cosino.io/product/5v-relays-array.

The BeagleBone Black's GPIOs and the pins of the relays array board I used with these devices are shown in the following table:

Pins	Relays array pin
P8.10 - GPIO68	IN1
P8.9 - GPIO69	IN2
P8.12 - GPIO44	IN3
P8.11 GPIO45	IN4
P9.1 - GND	GND
P9.6 - 5V	Vcc

To test the functionality of each GPIO line, we can use, as an example, the following command to enable one of them:

```
root@arm:~# ./bin/gpio_set.sh 68 out 1
```

 Note that the off state of the relay is 1, while the on state is 0.

Then, we can turn the relay on and off by just writing 0 and 1 in the /sys/class/gpio/gpio68/value file, as follows:

```
root@arm:~# echo 0 > /sys/class/gpio/gpio68/value
root@arm:~# echo 1 > /sys/class/gpio/gpio68/value
```

The final picture

The following image shows the prototype I realized to implement this project and to test the software. You should notice the infrared receiver on the bottom-right corner.

Note that, to use the relays array that needs the 5V power supply voltage, you must use an external power supply to power the BeagleBone Black.

Setting up the software

Now, it's time to set up the software to manage our infrared detector, and to do it, we're going to use the **LIRC (Linux Infrared Remote Control)** subsystem, which is a special code that has been developed for this purpose.

> For further information on the LIRC subsystem, you can take a look at http://www.lirc.org/.

We'll need a kernel driver to convert the pulse generated by the infrared detector into well-defined messages, and then to send them, through a LIRC device, to the userspace programs. At userspace level, we're going to use a special tool from the LIRC project in order to convert the infrared messages into input events, that is, the messages that a normal keyboard sends to the kernel.

 For further information on the Linux input subsystem, you can take a look at https://www.kernel.org/doc/Documentation/input/input.txt.

Setting up the kernel driver

To set up the kernel driver to manage our infrared receiver, we can use a procedure similar to the one used in *Chapter 4, Google Docs Weather Station*. Once the sources from the GitHub repository are downloaded, we have to follow the procedure in *Chapter 4, Google Docs Weather Station*, until the step where we have to apply our special patch. In fact, in this case, we have to apply the patch in the chapter_10/0001-Add-support-for-Homebrew-GPIO-Port-Receiver-Transmit.patch file in the book's example code repository in order to add the infrared receiver support.

The command is as follows:

```
$ git am --whitespace=nowarn 0001-Add-support-for-Homebrew-GPIO-Port-
Receiver-Transmit.patch
```

 Note that the --whitespace=nowarn command-line option is needed just in case your git system is configured to automatically fix up the whitespace errors, which is wrong in this case.

If everything works well, the git log command should display the following:

```
$ git log -1
commit be816108417ce82c7114ebd578ac32a45aef934a
Author:     Rodolfo Giometti <giometti@linux.it>
AuthorDate: Sun Oct 11 08:43:49 2015 +0200
Commit:     Rodolfo Giometti <giometti@linux.it>
CommitDate: Thu Oct 22 14:53:44 2015 +0200

    Add support for Homebrew GPIO Port Receiver/Transmitter

    Signed-off-by: Rodolfo Giometti <giometti@linux.it>
```

Before starting the kernel compilation, let me spend a few words regarding this patch. It simply adds a new driver into the KERNEL/drivers/staging/media/lirc directory of the Linux sources, so, after applying the patch, if we take a look at the new file lirc_gpio.c, we can discover how it works.

 The following is a brief explanation of the driver code. If you don't care about it, and just wish to use the driver *as-is*, you can safely skip this part until the kernel compilation commands.

At the beginning, we have the kernel module parameters as follows:

```
/*
 * Module parameters
 */

/* Set the default GPIO input pin */
staticintgpio_in = -1;
MODULE_PARM_DESC(gpio_in, "GPIO input/receiver pin number "
                "(warning: it MUST be an interrupt capable pin!)");
module_param(gpio_in, int, S_IRUGO);

/* Set the default GPIO output pin */
staticintgpio_out = -1;
MODULE_PARM_DESC(gpio_out, "GPIO output/transmitter pin number");
module_param(gpio_out, int, S_IRUGO);

/* Set the sense mode: -1 = auto, 0 = active high, 1 = active low */
staticint sense = -1;
MODULE_PARM_DESC(sense,
    "Override autodetection of IR receiver circuit: "
    "0 = active high, 1 = active low (default -1 = auto)");
module_param(sense, int, S_IRUGO);

/* Use softcarrier by default */
static unsigned int softcarrier = 1;
MODULE_PARM_DESC(softcarrier,
    "Software carrier: 0 = off, 1 = on (default on)");
module_param(softcarrier, uint, S_IRUGO);
```

We're going to use the `gpio_in` parameter in order to specify the input pin which the infrared receiver is connected to. Then, some local functions follow (which I don't report here), and then we find the file's operations definitions:

```
static const struct file_operationslirc_fops = {
        .owner          = THIS_MODULE,
        .write          = lirc_write,
        .unlocked_ioctl = lirc_ioctl,
        .read           = lirc_dev_fop_read,
        .poll           = lirc_dev_fop_poll,
        .open           = lirc_dev_fop_open,
        .release        = lirc_dev_fop_close,
        .llseek         = no_llseek,
};
```

Each function is related to a well-defined system call that we can use on the new LIRC device.

At the very bottom of the file, there is the `lirc_gpio_init_module()` function, which is responsible for setting up the new device. As a first step, this function tries to request all needed GPIO lines:

```
/*
 * Check for valid gpio pin numbers
 */
ret = gpio_request(gpio_in, LIRC_GPIO_NAME " ir/in");
if (ret) {
   pr_err("failed to request GPIO %u\n", gpio_in);
   return -EINVAL;
}
ret = gpio_direction_input(gpio_in);
if (ret) {
   pr_err("failed to set pin direction for gpio_in\n");
   ret = -EINVAL;
   goto exit_free_gpio_in;
}
pr_info("got GPIO %d for receiving\n", gpio_in);
/* Is GPIO in pin IRQ capable? */
irq = gpio_to_irq(gpio_in);
if (irq < 0) {
   pr_err("failed to map GPIO %d to IRQ\n", gpio_in);
   ret = -EINVAL;
```

```
        goto exit_free_gpio_in;
    }
    ret = request_irq(irq, (irq_handler_t) irq_handler,
        IRQF_TRIGGER_FALLING | IRQF_TRIGGER_ RISING,
        LIRC_GPIO_NAME, (void *) 0);
    if (ret < 0) {
        pr_err("unable to request IRQ %d\n", irq);
        goto exit_free_gpio_in;
    }
    pr_info("got IRQ %d for GPIO %d\n", irq, gpio_in);
    if (gpio_out >= 0) {
        ret = gpio_request(gpio_out, LIRC_GPIO_NAME " ir/ out");
        if (ret) {
            pr_err("failed to request GPIO %u\n", gpio_ out);
            goto exit_free_irq;
        }
        ret = gpio_direction_output(gpio_out, 0);
        if (ret) {
            pr_err("failed to set pin direction for gpio_ out\n");
            ret = -EINVAL;
            goto exit_free_gpio_out;
        }
        pr_info("got GPIO %d for transmitting\n", gpio_out);
    }
```

After requesting the `gpio_in` pin, the function sets it up as an input pin and then checks if such a GPIO line is interrupt-capable; otherwise, the driver can't work properly. If so, the function requests the IRQ line, and then it proceeds with the `gpio_out` pin (note that it's not mandatory).

Then, the driver sets the sense mode by using a little auto-detect procedure (if not directly specified by the user at loading time), as shown in the following code snippet:

```
        /* Set the sense mode */
        if (sense != -1) {
                pr_info("manually using active %s receiver on GPIO %d\n",
```

```
                        sense ? "low" : "high", gpio_in);
        } else {
                /* wait 1/2 sec for the power supply */
                msleep(500);

                /*
                 * probe 9 times every 0.04s, collect "votes" for
                 * active high/low
                 */
                nlow = 0;
                nhigh = 0;
                for (i = 0; i < 9; i++) {
                        if (gpio_get_value(gpio_in))
                                nlow++;
                        else
                                nhigh++;
                        msleep(40);
                }
                sense = (nlow >= nhigh ? 1 : 0);
                pr_info("auto-detected active %s receiver on GPIO pin
%d\n",
                        sense ? "low" : "high", gpio_in);
        }
```

Then, we can finally set up the LIRC driver by calling first the `lirc_buffer_init()` function, to properly allocate a memory buffer for the messages management, and then by calling the `lirc_register_driver()`, to register the driver into the system, as shown in the following code snippet:

```
        /*
         * Setup the LIRC driver
         */

        ret = lirc_buffer_init(&rbuf, sizeof(int), RBUF_LEN);
        if (ret < 0) {
            pr_err("unable to init lirc buffer!\n");
                ret = -ENOMEM;
                goto exit_free_gpio_out;
        }

        ret = platform_driver_register(&lirc_gpio_driver);
        if (ret) {
                pr_err("error in lirc register\n");
                goto exit_free_buffer;
```

```
        }

        lirc_gpio_dev = platform_device_alloc(LIRC_GPIO_NAME, 0);
        if (!lirc_gpio_dev) {
                pr_err("error on platform device alloc!\n");
                ret = -ENOMEM;
goto exit_driver_unreg;
        }

        ret = platform_device_add(lirc_gpio_dev);
        if (ret) {
                pr_err("error on platform device add!\n");
goto exit_device_put;
        }

        driver.features = LIRC_CAN_REC_MODE2;
        if (gpio_out >= 0) {
                driver.features |= LIRC_CAN_SET_SEND_DUTY_CYCLE |
                        LIRC_CAN_SET_SEND_CARRIER |
                        LIRC_CAN_SEND_PULSE;
        }

        driver.dev = &lirc_gpio_dev->dev;
        driver.minor = lirc_register_driver(&driver);

        if (driver.minor < 0) {
                pr_err("device registration failed!");
                ret = -EIO;
goto exit_device_put;
        }

        pr_info("driver registered!\n");

        return 0;
```

Ok, now we can start to compile the kernel with the following command:

```
$ ./build_kernel.sh
```

 This step, and the subsequent one, are time consuming and require patience, so you should take a cup of your preferred tea or coffee, and just wait.

After some time, the procedure will present the standard kernel configuration panel, and now we should verify that the needed drivers are enabled. You should navigate in the menu to **Device Drivers | Staging drivers | Media staging drivers | Linux Infrared Remote Control IR receiver/transmitter drivers**, where the **Homebrew GPIO Port Receiver/Transmitter** entry should be selected as module (**<M>**).

Then, exit the configuration menu and the kernel compilation should start. Then, when it ends, the new kernel image will be ready, and the following message should appear:

```
- - - - - - - - - - - - - - - - - - - - - - - - - - -
Script Complete
eewiki.net: [user@localhost:~$ export kernel_version=3.13.11-bone12]
- - - - - - - - - - - - - - - - - - - - - - - - - - -
```

Now, we can install it on the microSD, using the following installation tool:

```
$ ./tools/install_kernel.sh
```

If everything works well, after the usual login, we can verify that the new kernel is really running using the following command:

```
root@arm:~# uname -a
Linux arm 3.13.11-bone12 #1 SMP Sun Oct 11 09:15:46 CEST 2015 armv7l GNU/
Linux
```

 Note that the kernel version on your system may be more recent than mine.

Okay, the new kernel is ready! Now, we can load the LIRC driver by using the following command:

```
root@arm:~# modprobe lirc_gpio gpio_in=60
```

Note that the GPIO 60 must not be in use, or you may get an error like the following:
```
ERROR: could not insert 'lirc_gpio': Invalid argument
```

The kernel messages should look like the following:

```
lirc_dev: IR Remote Control driver registered, major 241
lirc_gpio: module is from the staging directory, the quality is unknown,
you have been warned.
lirc_gpio: got GPIO 60 for receiving
```

```
lirc_gpio: got IRQ 204 for GPIO 60
lirc_gpio: auto-detected active low receiver on GPIO pin 60
lirc_gpio lirc_gpio.0: lirc_dev: driver lirc_gpio registered at minor = 0
lirc_gpio: driver registered!
```

Also, a new entry should now be ready under the /dev directory:

```
root@arm:~/chapter_10# ls -l /dev/lirc*
crw-rw---T 1 root video 241, 0 Aug 13 16:35 /dev/lirc0
```

The LIRC tools

Now that the kernel module is set up and running, we need some userspace tools to manage it. So, let's install the lirc package with the usual aptitude command:

```
root@arm:~# aptitude install lirc
...
Setting up lirc (0.9.0~pre1-1) ...
[ ok ] No valid /etc/lirc/lircd.conf has been found..
[ ok ] Remote control support has been disabled..
[ ok ] Reconfigure LIRC or manually replace /etc/lirc/lircd.conf to
enable..
```

As stated by the preceding line, to enable the lircd daemon (that is, the tools we need), we have to replace the configuration file /etc/lirc/lircd.conf; however, we're not going to use the daemon this way. In reality, we can test that the driver is really working as expected by executing the following command:

```
root@arm:~# mode2 --driver default --device /dev/lirc0
```

Nothing should happen until you point your remote controller at the infrared receiver and press a button. In this case, you should see some output, as follows:

```
space 3333126
pulse 8985
space 4503
pulse 564
space 535
pulse 564
space 561
pulse 542
space 551
...
```

Okay! The /dev/lirc0 device is functioning, and the driver correctly detects the messages from the remote controller! Now, we have to create a custom configuration file in order to associate an input event to each of the remote controller's buttons.

 Again, due to lack of space, I'm going to configure just a few buttons in the following example; but you can add whatever you want.

The command to use is irrecord, as follows:

```
root@arm:~# irrecord --driver default --device /dev/lirc0 myremote.conf
```

myremote.conf is the file where we wish to save our configuration. The program then will show an output as follows:

```
irrecord -  application for recording IR-codes for usage with lirc

Copyright (C) 1998,1999 Christoph Bartelmus(lirc@bartelmus.de)


This program will record the signals from your remote control
and create a config file for lircd.



A proper config file for lircd is maybe the most vital part of this
package, so you should invest some time to create a working config
file. Although I put a good deal of effort in this program it is often
notpossible to automatically recognize all features of a remote
control. Often short-comings of the receiver hardware make it nearly
impossible. If you have problems to create a config file READ THE
DOCUMENTATION of this package, especially section "Adding new remote
controls" for how to get help.

If there already is a remote control of the same brand available at
http://www.lirc.org/remotes/ you might also want to try using such a
remote as a template. The config files already contain all
parameters of the protocol used by remotes of a certain brand and
knowing these parameters makes the job of this program much
easier. There are also template files for the most common protocols
available in the remotes/generic/ directory of the source
```

```
distribution of this package. You can use a template files by
providing the path of the file as command line parameter.

Please send the finished config files to <lirc@bartelmus.de> so that I
can make them available to others. Don't forget to put all information
that you can get about the remote control in the header of the file.

Press RETURN to continue.
```

Okay, let's press the *return/Enter* key, and the program will continue showing the following message:

```
Now start pressing buttons on your remote control.

It is very important that you press many different buttons and hold them
down for approximately one second. Each button should generate at least
one dot but in no case more than ten dots of output.
Don't stop pressing buttons until two lines of dots (2x80) have been
generated.

Press RETURN now to start recording.
```

Ok, now it's really important to carefully follow the preceding instructions. So, let's start pressing different buttons and hold them for approximately one second in such a way as to generate at least one dot, but in no case more than ten dots, of output for each press!

So, the program will start printing the dots until it reaches the end of the terminal as follows:

```
. . . . . . . . . . . . . . . . . . . . . . . . . . . . . . . . . . . . . . . . . . . . . . . . . . . . . . . . . . . . . . . . . . . . . . . . .
Found const length: 107736
```

When the first line is finished, the program will display the following message, and new dots will appear, but, this time, just one per pressed button!:

```
Please keep on pressing buttons like described above.
...........irrecord: signal too long
Creating config file in raw mode.
Now enter the names for the buttons.
```

Now, the first detection stage is finished, and we can start the real detection one button at a time. Now, the system will ask for a button name or the *Enter* key to finish:

```
Please enter the name for the next button (press <ENTER> to finish
recording)
```

Now, I enter the name of button **0** by inserting the KEY_0 string, as follows. Then, the system will ask you to hold down the button **0** until it has got it:

```
KEY_0


Now hold down button "KEY_0".

Got it.

Signal length is 67
```

> The valid button names can be listed by using the irrecord command,
> as follows:
> ```
> root@arm:~# irrecord --list-namespace
> KEY_0
> KEY_102ND
> KEY_1
> KEY_2
> KEY_3
> KEY_4
> KEY_5
> KEY_6
> KEY_7
> KEY_8
> KEY_9
> KEY_A
> KEY_AB
> ...
> ```

Then, the procedure restarts for the next buttons as follows:

```
Please enter the name for the next button (press <ENTER> to finish
recording)
KEY_1


Now hold down button "KEY_1".

Got it.
```

```
Signal length is 67


Please enter the name for the next button (press <ENTER> to finish
recording)
KEY_2


Now hold down button "KEY_2".
Got it.
Signal length is 67


Please enter the name for the next button (press <ENTER> to finish
recording)
KEY_3


Now hold down button "KEY_3".
Got it.
Signal length is 67
```

At this point, I simply enter no names and just press *Enter* to exit, and then I get the prompt again:

```
Please enter the name for the next button (press <ENTER> to finish
recording)
```

```
root@arm:~#
```

Now, a new file called `myremote.conf` should be ready. The following is a snippet of my file:

```
# Please make this file available to others
# by sending it to <lirc@bartelmus.de>
#
# this config file was automatically generated
# using lirc-0.9.0-pre1(default) on Wed Aug 13 15:54:26 2014
#
# contributed by
#
# brand:                        myremote.conf
# model no. of remote control:
# devices being controlled by this remote:
#
```

```
begin remote

  name  myremote.conf
  flags RAW_CODES
  eps          30
  aeps        100

  gap       96036

begin raw_codes

        name KEY_0
          8998    4478     566     541     570     541
           570     541     570     542     570     541
           570     541     570     541     578     533
           570     541     570     541     570     542
           570     540     570    1679     571     541
           570     541     569     543     569     542
           570     541     570    1679     570    1678
           571     541     570     541     570     542
           570     540     570    1679     570    1679
           570     541     571     540     571    1685
           563    1679     570    1678     571    1678
           571   47910    9003    2231     570

        name KEY_1
          8969    4507     537     571     539     571
           540     572     539     572     539     572
           540     571     540     572     546     565
           539     572     540     571     540     571
           540     571     540    1709     540     572
           539     572     546     566     538    1709
           540    1709     540     572     539     572
           539     572     539     573     545     566
           538     572     539     573     538     572
           539    1710     540    1709     539    1712
           539    1709     539    1709     539    1710
           539   47930    8983    2261     539

  ...
```

Now, we are ready to test our job. We have to verify if all buttons have been correctly recognized. To do it, we have to execute the `lircd` daemon from the command line as follows:

```
root@arm:~# lircd --nodaemon --device /dev/lirc0 --driver default
--uinput myremote.conf
lircd-0.9.0-pre1[2235]: lircd(default) ready, using /var/run/lirc/lircd
```

The last argument `--uinput` is used to instruct the `lircd` daemon to convert the button presses into input events as they came from a normal keyboard, so, we can test them with the `evtest` command. It must be executed into another terminal due to the fact that the previous command must run with `evtest` at the same time! The command is as follows:

```
root@arm:~# evtest
No device specified, trying to scan all of /dev/input/event*
Available devices:
/dev/input/event0:      lircd
Select the device event number [0-0]:
```

Now, we have to select the (only) available input device with the `0` number, and the program will continue showing the following output:

```
Input driver version is 1.0.1
Input device ID: bus 0x0 vendor 0x0 product 0x0 version 0x0
Input device name: "lircd"
Supported events:
  Event type 0 (EV_SYN)
  Event type 1 (EV_KEY)
    Event code 1 (KEY_ESC)
    Event code 2 (KEY_1)
    Event code 3 (KEY_2)
    Event code 4 (KEY_3)
    ...
    Event code 237 (KEY_BLUETOOTH)
    Event code 238 (KEY_WLAN)
    Event code 239 (KEY_UWB)
    Event code 240 (KEY_UNKNOWN)
  Event type 20 (EV_REP)
Properties:
Testing ... (interrupt to exit)
```

Then, when I press a button on my remote controller, I get the following output:

```
Event: time 1445765562.506427, type 1 (EV_KEY), code 11 (KEY_0), value 1
Event: time 1445765562.506427, ------------- SYN_REPORT -----------
...
Event: time 1445765566.745716, type 1 (EV_KEY), code 2 (KEY_1), value 1
Event: time 1445765566.745716, ------------- SYN_REPORT -----------
...
Event: time 1445765568.216621, type 1 (EV_KEY), code 3 (KEY_2), value 1
Event: time 1445765568.216621, ------------- SYN_REPORT -----------
...
Event: time 1445765569.357041, type 1 (EV_KEY), code 4 (KEY_3), value 1
Event: time 1445765569.357041, ------------- SYN_REPORT -----------
...
```

 The evtest program can be installed by using the following command:

```
root@arm:~# aptitude install evtest
```

Note that the **0** button on the remote controller corresponds to the KEY_0 input event that has the 11 code, while the **1**, **2**, and **3** buttons correspond to the KEY_1, KEY_2, and KEY_3 input events that have the 2, 3, and 4 codes. So, we can map such events with the corresponding GPIO line by using a look-up table as follows (Python syntax):

```
GPIO = [-1, -1, 69, 44, 45, -1, -1, -1, -1, -1, -1, 68]
```

The -1 value means *no GPIO*. So, when we press the **0** button, we receive the KEY_0 input event that has the 11 code, and at the 11th position of the array (starting the count from 0), we have the 68 value, so, the GPIO68 is attached to the **0** button on the remote controller. In a similar manner, the **1**, **2**, and **3** buttons that correspond to the KEY_1 (code 2), KEY_2 (code 3), and KEY_3 (code 4) input events, are connected to GPIO 69 (array index 2), GPIO 44 (array index 3), and GPIO 45 (array index 4) respectively.

The input events manager

Now, we have to add the last element only; that is, the software that takes the input events and turns the corresponding relay on and off. To do it in a *dirty and quick* way, we can use the Python language with the evdev library that can be easily installed on our BeagleBone Black with the following command:

```
root@arm:~# pip install evdev
```

 The curious reader can get more information about this library at https://python-evdev.readthedocs.org/ en/latest/.

After the library has been installed, we can consider a possible implementation of our input events manager, as shown in the following code snippet:

```
#
# Local functions
#

def gpio_get(gpio):
    fd = open("/sys/class/gpio/gpio" + str(gpio) + "/value", "r")
    val = fd.read()
    fd.close()
return int(val)

def gpio_set(gpio, val):
    fd = open("/sys/class/gpio/gpio" + str(gpio) + "/value", "w")
    v = fd.write(str(val))
    fd.close()

def usage():
    print("usage: ", NAME, " [-h] <inputdev>", file=sys.stderr)
    sys.exit(2);

#
# Main
#

try:
    opts, args = getopt.getopt(sys.argv[1:], "h",
        ["help"])
except getopt.GetoptError, err:
    # Print help information and exit:
    print(str(err), file=sys.stderr)
```

```
         usage()

for o, a in opts:
   if o in ("-h", "--help"):
      usage()
   else:
      assert False, "unhandled option"

# Check command line
if len(args) < 1:
   usage()

# Try to open the input device
try:
   dev = InputDevice(args[0])
except:
   print("invalid input device", args[0], file=sys.stderr)
   sys.exit(1);

logging.info (dev)
logging.info("hit CTRL+C to stop")

# Start the main loop
for event in dev.read_loop():
    if event.type == ecodes.EV_KEY and event.value == 1:
        # Get the key code and convert it to the corresponding GPIO
        code = event.code
        if code < 0 or code > len(GPIO):
                gpio = -1
        else:
                gpio = GPIO[code]
        logging.info("got code %d -> GPIO%d" % (code, gpio))

        if gpio > 0:
                # Get current GPIO status and invert it
                status = gpio_get(gpio)
                status = 1 - status
                gpio_set(gpio, status)
                logging.info("turning GPIO%d %d -> %d" %
                        (gpio, 1 - status, status))
        else:
                logging.info("invalid button")
```

 The complete code is stored in the chapter_10/read_events.
py file in the book's example code repository.

The code is quite self explanatory, but let me explain some points. First of all, note that the GPIO array is the one defined in the previous section, and then the gpio_ get() and gpio_set() methods are used to get and set a GPIO status. The program, after a little check to the command line, starts opening the input device supplied by the user by using the InputDevice() method and then enters into the big loop, where it waits for a key press, and then it switches the status of the corresponding GPIO (if any).

The following is a sample usage:

```
root@arm:~# ./read_events.py /dev/input/event0
INFO:root:device /dev/input/event0, name "lircd", phys ""
INFO:root:hit CTRL+C to stop
INFO:root:got code 2 -> GPIO68
INFO:root:turning GPIO68 1 -> 0
INFO:root:got code 3 -> GPIO69
INFO:root:turning GPIO69 1 -> 0
INFO:root:got code 3 -> GPIO69
INFO:root:turning GPIO69 0 -> 1
INFO:root:got code 2 -> GPIO68
INFO:root:turning GPIO68 0 -> 1
```

Now, before continuing, let me suggest to you an interesting feature in using Linux's input layer.

Even if it may seem a bit complicated using the input layer instead of directly accessing the lircd daemon, this approach has the big advantage that we can test our relays manager with any input device! In fact, if you try to connect a normal keyboard to the BeagleBone Black's USB port, you'll get a new input device as follows:

```
root@arm:~# evtest
No device specified, trying to scan all of /dev/input/event*
Available devices:
/dev/input/event0:    lircd
/dev/input/event1:    HID 04d9:1203
Select the device event number [0-1]:
```

Now, selecting the new input device /dev/input/event1, we can generate the same input events as before by simply pressing the **0, 1, 2**, and **3** keys:

```
Event: time 1445766356.367407, type 4 (EV_MSC), code 4 (MSC_SCAN), value
70027
Event: time 1445766356.367407, type 1 (EV_KEY), code 11 (KEY_0), value 1
Event: time 1445766356.367407, -------------- SYN_REPORT -----------
...
Event: time 1445766365.537391, type 4 (EV_MSC), code 4 (MSC_SCAN), value
7001e
Event: time 1445766365.537391, type 1 (EV_KEY), code 2 (KEY_1), value 1
Event: time 1445766365.537391, -------------- SYN_REPORT -----------
...
Event: time 1445766367.437377, type 4 (EV_MSC), code 4 (MSC_SCAN), value
7001f
Event: time 1445766367.437377, type 1 (EV_KEY), code 3 (KEY_2), value 1
Event: time 1445766367.437377, -------------- SYN_REPORT -----------
...
Event: time 1445766369.537383, type 4 (EV_MSC), code 4 (MSC_SCAN), value
70020
Event: time 1445766369.537383, type 1 (EV_KEY), code 4 (KEY_3), value 1
Event: time 1445766369.537383, -------------- SYN_REPORT -----------
...
```

 Note that, even if not shown here due to lack of space, the keyboard generates more input events than the usual EV_KEY ones. But we can easily skip them just by selecting the right input event type.

In this situation, if we execute our program as in the following command line, we can manage the relays as we did with the remote controller:

```
root@arm:~# ./read_events.py /dev/input/event1
INFO:root:device /dev/input/event1, name "HID 04d9:1203", phys "usb-musb-
hdrc.1.auto-1/input0"
INFO:root:hit CTRL+C to stop
INFO:root:got code 11 -> GPIO68
INFO:root:turning GPIO68 1 -> 0
INFO:root:got code 2 -> GPIO69
INFO:root:turning GPIO69 1 -> 0
```

```
INFO:root:got code 3 -> GPIO44
INFO:root:turning GPIO44 1 -> 0
INFO:root:got code 4 -> GPIO45
INFO:root:turning GPIO45 1 -> 0
```

The final test

As in the previous chapters, we have to first execute the chapter_10/SYSINIT.sh file in the book's example code repository as usual to set up all GPIOs lines and to load the kernel module:

```
root@beaglebone:~# ./SYSINIT.sh
done!
```

Then, we must execute the lircd daemon by using the command line without the --nodaemon option argument:

```
root@arm:~# lircd --device /dev/lirc0 --driver default --uinputmyremote.conf
```

Then, we can execute the preceding read_events.py program to manage the relays:

```
root@arm:~# ./read_events.py /dev/input/event0
```

Now the trick is done. We simply have to direct the remote controller to the infrared detector and press the **0**, **1**, **2**, or **3** button. When we press the button, the switch turns on, while when we press the button again, the switch also turns off changing the status of the connected relay, and, as a last step, the device connected to it.

Summary

In this chapter, we took a look at a kernel driver to manage an infrared device. Then, we saw how to use the userspace tools from the LIRC project to receive the messages from the remote controller, and then to turn them into specific Linux input events. This allowed us to manage some devices connected with our BeagleBone Black.

In the next chapter, we'll discover how to manage a wireless device to manage a wall plug and to monitor the power consumption of the device connected to it by using a common communication system for the home automation, that is, the **Z-Wave** protocol.

11
A Wireless Home Controller with Z-Wave

In this project, we'll see how to implement a little wireless home controller by using a Z-Wave controller connected with our BeagleBone Black and two Z-Wave devices: a wall plug and a multisensor device. With the former, we'll be able to turn on and off every household appliance connected to it, and, at the same time, measure its power consumption. With the latter, we'll be able to measure several environment variables like temperature, humidity, and luminance (and have a motion detector capability too).

The Z-Wave communication protocol allows us to manage several home automation sensors and actuators wirelessly, so we don't need to modify our pre-existing plant. Also, we can easily add a power consumption measuring system or several environment sensors with a minor impact on the actual home layout.

As a last step, to keep the code simple, but in order to allow the user to easily manage the system, we'll write a simple web interface written in Python to easily manage the prototype.

The basics of functioning

This time, the project is a bit more complex than before, but all the complexity is not in the hardware (the connections are very simple, just plug in a USB dongle and the trick is done!,) but in the software! In fact, the management software to set up and control these devices needs some skills. Also, due to the fact that the Z-Wave world is really huge, and the lack of space for it in this book (I suppose I can ask my editor to write a dedicated book just to explain how to use the Z-Wave for the home automation projects!), I'm just going to present the very basics of the Z-Wave protocol, showing a minimal application that you can expand on your own.

As stated before, we're going to use a Z-Wave controller connected to the BeagleBone Black's USB host port to manage two Z-Wave devices: one used to measure some environment data, and one to turn on and off a connected device. So, what we have to do is to write some software to be able to send and receive messages to/from these devices via the controller in order to exchange data and commands between the BeagleBone Black and the two slave devices. The code we're going to write should have a part to manage the Z-Wave messages and a part to interact with the user. Regarding this last part, I decided to use a web interface written in Python with a little code in HTML and JavaScript.

Setting up the hardware

The Z-Wave technology, which is oriented to the residential control and automation market, is designed to be suitable for battery-operated devices. In fact, one of its main goals is to minimize the power consumption. Despite this fact, it provides reliable and low-latency transmission of small data packets at data rates of up to 100 kbps, and a simple yet reliable method to wirelessly manage sensors and control lights and appliances in a house.

 For more information on Z-Wave, a good starting point is at `https://en.wikipedia.org/wiki/Z-Wave`.

In our project, we're going to use a Z-Wave controller on a USB dongle, one slave device powered by the same plug where it's connected to, and one multisensor device that can be powered by batteries or via an external USB connection.

Setting up the Z-Wave controller

The Z-Wave controller I used in this prototype is shown in the following image:

The device can be purchased at the following link (or by surfing the Internet): http://www.cosino.io/product/usb-z-wave-controller.

A reference design is available here:

http://z-wave.sigmadesigns.com/docs/brochures/UZB_br.pdf.

Once connected with the BeagleBone Black's USB host port by using the lsusb command, we should get the following output:

```
root@beaglebone:~# lsusb
Bus 001 Device 002: ID 0658:0200 Sigma Designs, Inc.
Bus 001 Device 001: ID 1d6b:0002 Linux Foundation 2.0 root hub
Bus 002 Device 001: ID 1d6b:0002 Linux Foundation 2.0 root hub
```

Also we should see the following kernel activity:

```
hub 1-0:1.0: hub_resume
hub 1-0:1.0: port 1: status 0101 change 0001
hub 1-0:1.0: state 7 ports 1 chg 0002 evt 0000
hub 1-0:1.0: port 1, status 0101, change 0000, 12 Mb/s
usb 1-1: new full-speed USB device number 2 using musb-hdrc
usb 1-1: ep0 maxpacket = 8
usb 1-1: skipped 4 descriptors after interface
usb 1-1: udev 2, busnum 1, minor = 1
usb 1-1: New USB device found, idVendor=0658, idProduct=0200
usb 1-1: New USB device strings: Mfr=0, Product=0, SerialNumber=0
usb 1-1: usb_probe_device
usb 1-1: configuration #1 chosen from 1 choice
usb 1-1: adding 1-1:1.0 (config #1, interface 0)
cdc_acm 1-1:1.0: usb_probe_interface
cdc_acm 1-1:1.0: usb_probe_interface - got id
cdc_acm 1-1:1.0: This device cannot do calls on its own. It is not a
modem.
cdc_acm 1-1:1.0: ttyACM0: USB ACM device
usb 1-1: adding 1-1:1.1 (config #1, interface 1)
hub 1-0:1.0: state 7 ports 1 chg 0000 evt 0002
hub 1-0:1.0: port 1 enable change, status 00000103
```

Looking at the last but fourth line, we can discover that the Z-Wave controller has been connected to the `/dev/ttyACM0` device file. So, the device is correctly connected. But to really test it, we need to install a proper management software. To do so, we can use an open source implementation of the Z-Wave protocol named **Open Z-Wave**, where we can find a lot of suitable software to test a Z-Wave network.

 The home page of the Open Z-Wave project is at `http://www.openzwave.com`.

With the following command, we can download the code we need into our prototype:

```
root@beaglebone:~# git clone https://github.com/OpenZWave/open-zwave
openzwave
```

Then, we need some extra packages to compile the needed tools. So, let's install them with the following command:

```
root@beaglebone:~# aptitude install build-essential make git libudev-dev
libjson0 libjson0-dev libcurl4-gnutls-dev
```

Now, just enter into the `openzwave` directory, and simply use the `make` command as follows:

```
root@beaglebone:~# cd openzwave
root@beaglebone:~/openzwave# make
```

 The compilation is quite slow, so be patient.

When finished, we need to download another repository into the current directory with the following command:

```
root@beaglebone:~/openzwave# git clone https://github.com/OpenZWave/open-zwave-control-panel openzwave-control-panel
```

Then, after the download, we have to install an extra package to proceed with the compilation. So, let's use the `aptitude` command again, as follows:

```
root@beaglebone:~/openzwave# aptitude install libmicrohttpd-dev
```

Now, as the last step, enter into the `openzwave-control-panel` directory and rerun the `make` command with the following command lines:

```
root@beaglebone:~/openzwave# cd openzwave-control-panel/
root@beaglebone:~/openzwave/openzwave-control-panel# make
```

When the compilation is finished, the `ozwcp` program should be available. So, let's execute it by using the following command lines:

```
root@beaglebone:~/openzwave/openzwave-control-panel# ln -s ../config
root@beaglebone:~/openzwave/openzwave-control-panel# ./ozwcp -d -p 8080
2014-04-23 21:12:52.943 Always, OpenZwave Version 1.3.526 Starting Up
webserver starting port 8080
```

Note that the `ln` command is just used once to create a proper link with the Open Z-Wave configuration directory `config`, which is located in the upper directory.

If you get the following error while executing the program, it means that most probably your web server is holding port `8080`, so you have to disable it:

```
Failed to bind to port 8080: Address already in use
```

Now, we should point the web browser on our host PC to the address `http://192.168.7.2:8080/` to get what is shown in the following screenshot:

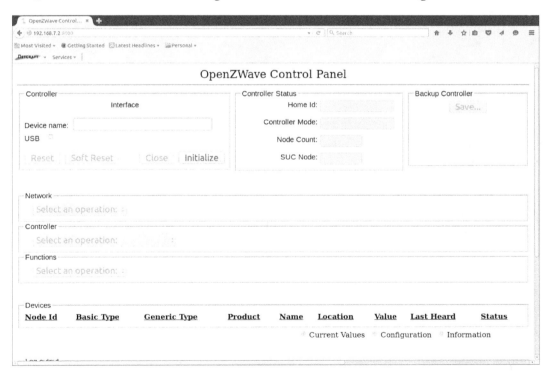

Okay, now we have to enter the `/dev/ttyACM0` path name into the **Device name** field, and then press the **Initialize** button to start the communication. If everything works well, you should see that a new device is listed in the **Devices** tab, as shown in the following screenshot:

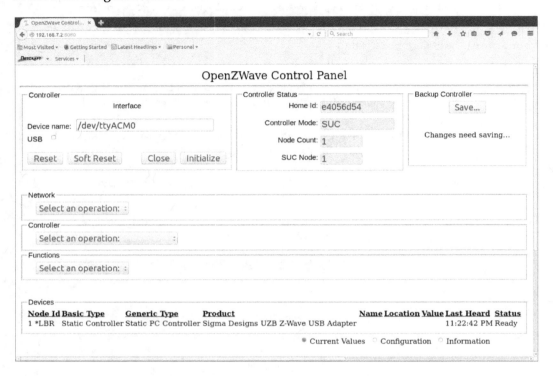

Now, the controller is up and running, so, we can continue installing the Z-Wave slaves.

Setting up the Z-Wave wall plug

The first Z-Wave slave is the wall plug shown in the following image:

The device can be purchased at the following link (or by surfing the Internet): http://www.cosino.io/product/z-wave-wall-plug.

A reference manual is available here:

http://www.fibaro.com/manuals/en/FGWPx-101/FGWPx-101-EN-A-v1.00.pdf.

The device is wireless and, once connected with a powered plug, it's self-powered; so, we don't need special connections to set it up. However, we need some home appliance connected to it, as shown in the following image, for the power consumption measurements:

Now, to test this device and its communication with the controller, we can use the ozwcp program again. Just click on the **Select an operation** menu entry in the **Controller** tab and select the **Add Device** entry and then press the **Go** button. On the left, you should see the **Add Device: waiting for a user action** message. So, let's power up the device by putting it into a wall plug and then strike the button on the device in order to start the pairing procedure (just as a Bluetooth device does).

 Note that newer version of this device doesn't require you to press the button to start the pairing procedure—it just starts automatically after the first plug.

If everything works well, a new device should appear in the **Devices** tab as follows:

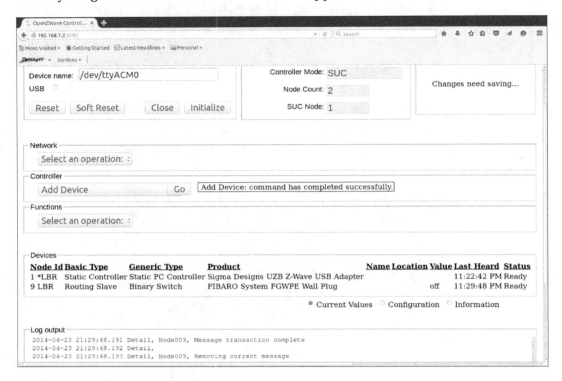

Now, we can change some of the device's settings by selecting the new device and then clicking on the **Configuration** option under the **Devices** listing tab. A panel setting similar to the following screenshot should appear:

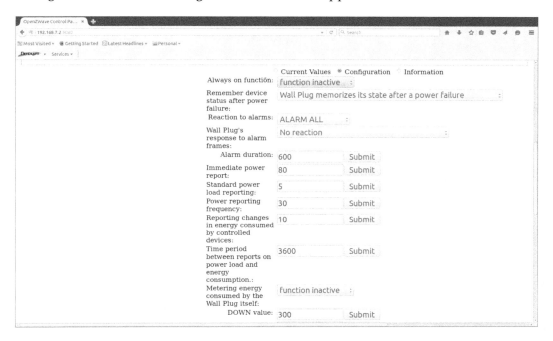

Now, we can change the **Standard power load reporting** entry by writing the new value in the related field and then pressing the **Submit** button. In this manner, we can define a lower value by how much power load must change (in percentage) to be reported to the main controller (I used the value 5).

Setting up the Z-Wave multisensor

The second Z-Wave slave is the multisensor shown in the following image:

> The device can be purchased at the following link (or by surfing the Internet): http://www.cosino.io/product/z-wave-multi-sensor.
>
> A reference manual is available here:
>
> http://aeotec.com/z-wave-sensor/47-multisensor-manual.html.

To power the device, we can use 4 batteries or a USB cable connected as in the following image. Then, to test the device and its communication with the controller, we can use the ozwcp program again. So, just click on the **Select an operation** menu entry in the **Controller** tab and select the **Add Device** entry. Then, press the **Go** button in order to repeat a pairing procedure again (the pairing button is the black button near the sensitivity regulator under the battery pack cover pack cover. In the image below it is located in the top-right corner).

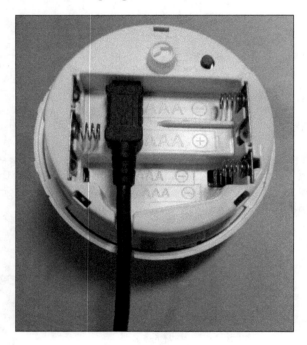

Again, if everything works well, a new device should appear in the **Devices** tab as follows:

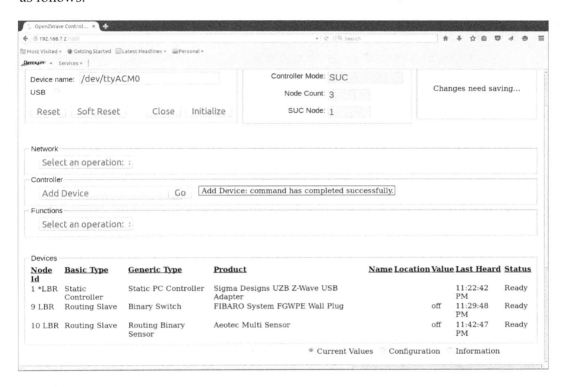

Now, as before, we can change the default settings. In particular, we can set up the environment report's frequencies and the report's content by changing the **Group 1 Reports** entry to 224 and the **Group 2 Reports** entry to 1, then **Group 1 Interval** to 10 and **Group 2 Interval** to 60.

These special settings will instruct the multisensor to enable bits 7 (luminosity), 6 (humidity), and 5 (temperature) into group 1, and bit 0 (battery level) into group 2, and to repeat them every 10 seconds for group 1 and every 60 seconds for group 2 (see the following screenshot):

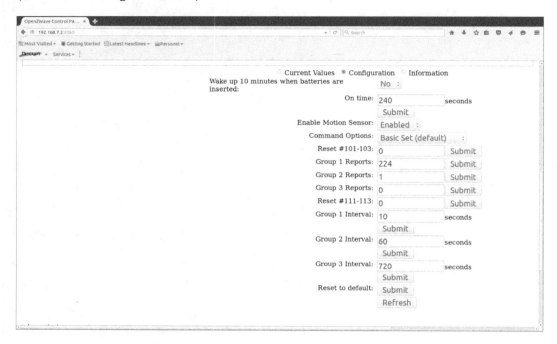

Ok, now all the devices are ready to operate! So, we can stop the `ozwcp` program by pressing the *CTRL + C* key sequence and go forward to the next sections.

The final picture

The following is the image that shows the prototype I realized to implement this project and to test the software.

Nothing special to say here; just the BeagleBone Black with the Z-Wave controller USB dongle and the two Z-Wave devices described before.

Setting up the software

As already stated, the more complex part of this prototype is the software. We need to install several software packages into our BeagleBone Black, and the software we have to write by ourselves needs some skills. However, don't worry, I'm going to explain all needed steps one at a time!

Setting up the Python bindings

Installing the **Python** bindings is quite complex since the software package named `python-openzwave` still seems in hard development, and it depends on tons of **Python** packages! However, I did it by getting a specific version of the project with the following command line:

```
root@beaglebone:~# wget http://bibi21000.no-ip.biz/python-openzwave/
python-openzwave-0.3.0b5.tgz
```

 Other versions of the `python-openzwave` package are available at
`http://bibi21000.no-ip.biz/python-openzwave/`.

Now, to explore the archive file, we can use the following command:

root@beaglebone:~# tar xvfz python-openzwave-0.3.0b5.tgz

A new directory, python-openzwave-0.3.0b5, is now created; however, to successfully compile the code, we need to install some Python packages by using the following command line several times:

root@beaglebone:~# pip install **<package>**

Here, with **<package>**, I used the following names: Louie, urwid, Flask-SocketIO, versiontools, gevent-socketio, WebOb, Flask-Themes, and Flask-Babel.

In reality, the pip install command accepts more packages at once separated by spaces, so you can use a single command to install all the needed packages at once.

Note also that to install the urwid package, I needed to use a different command to update the already installed package. The command is as follows:

root@beaglebone:~# pip install --upgrade urwid

Meanwhile, to install the Flask-WTF package at version 0.9.5, the command I used is as follows:

root@beaglebone:~# pip install Flask-WTF==0.9.5

However, on my system I cannot successfully execute it, so I used a *dirty-trick* by applying the following patch to the setup-web.py file in the python-openzwave-0.3.0b5 directory:

```
--- ./setup-web.py.orig    2014-04-24 03:50:41.398440723
+0000
+++ ./setup-web.py    2014-04-24 03:39:52.212893771
+0000
@@ -49,7 +49,7 @@
    install_requires = [
                        'openzwave == %s' % pyozw_
version,
                        'Flask == 0.10.1',
-                       'Flask-WTF == 0.9.5',
+                       'Flask-WTF >= 0.9.5',
                        'Babel >= 1.0',
                        'Flask-Babel == 0.9',
                        #'Flask-Fanstatic == 0.2.0',
```

Luckily, the code works correctly, even when using a newer version than 0.9.5!

After all dependencies have been installed, just move to the `python-openzwave-0.3.0b5` directory and use the `make` command to do the job, as follows:

```
root@beaglebone:~/python-openzwave-0.3.0b5# make deps build
```

 Again, the compilation is quite slow, so be patient!

When finished, we have to install the new code by using the following command line:

```
root@beaglebone:~/python-openzwave-0.3.0b5# make install
```

We did it! To test our new code, we can now use the provided examples by going into the `examples` directory, and then executing the following command lines:

```
root@beaglebone:~/python-openzwave-0.3.0b5# cd examples/
root@beaglebone:~/python-openzwave-0.3.0b5/examples# ./test_lib.py
--device=/dev/ttyACM0
```

Note that, for better readability in the output lines of the `test_lib.py` command, all timing references have been removed.

Also, for all the next Python codes, you can safely ignore all the warning messages in the following form:

```
./test_lib.py:28: UserWarning: Module libopenzwave was
already imported from None, but /usr/local/lib/python2.7/
dist-packages/libopenzwave-0.3.0b5-py2.7-linux-armv7l.egg
is being added to sys.path
```

The output of the preceding command is really long and, due to lack of space, I cannot report it here completely, so I'm going to report only the relevant parts.

In the first lines, we get some basic information messages, as follows:

```
Always, OpenZwave Version 1.3.482 Starting Up
Add watcher
Add device
Info, Setting Up Provided Network Key for Secure Communications
Warning, Failed - Network Key Not Set
Info, mgr,    Added driver for controller /dev/ttyACM0
Sniff network during 60.0 seconds
Info,   Opening controller /dev/ttyACM0
Info, Trying to open serial port /dev/ttyACM0 (attempt 1)
Info, Serial port /dev/ttyACM0 opened (attempt 1)
```

Here, there is some information regarding the software release and the device we're getting access to (that is /dev/ttyACM0). Then, there is a list of the queued commands to be executed, as follows:

```
Detail, contrlr, Queuing (Command) FUNC_ID_ZW_GET_VERSION: 0x01, 0x03,
0x00, 0x15, 0xe9

Detail, contrlr, Queuing (Command) FUNC_ID_ZW_MEMORY_GET_ID: 0x01, 0x03,
0x00, 0x20, 0xdc

Detail, contrlr, Queuing (Command) FUNC_ID_ZW_GET_CONTROLLER_
CAPABILITIES: 0x01, 0x03, 0x00, 0x05, 0xf9

Detail, contrlr, Queuing (Command) FUNC_ID_SERIAL_API_GET_CAPABILITIES:
0x01, 0x03, 0x00, 0x07, 0xfb

Detail, contrlr, Queuing (Command) FUNC_ID_ZW_GET_SUC_NODE_ID: 0x01,
0x03, 0x00, 0x56, 0xaa

Detail, contrlr, Sending (Command) FUNC_ID_ZW_GET_VERSION: 0x01, 0x03,
0x00, 0x15, 0xe9

Detail, contrlr, Received: 0x01, 0x10, 0x01, 0x15, 0x5a, 0x2d, 0x57,
0x61, 0x76, 0x65, 0x20, 0x33, 0x2e, 0x37, 0x39, 0x00, 0x01, 0x9b
```

Here, are some first answers:

```
Info, contrlr,   Received reply to FUNC_ID_ZW_GET_VERSION:
Info, contrlr,   Static Controller library, version Z-Wave 3.79
```

Then, a lot of messages regarding the probing of all available **Z-Wave** nodes follow:

```
Info, contrlr,     Node 001 - New
Detail, Node001, AdvanceQueries queryPending=0 queryRetries=0
queryStage=None live=1
Detail, Node001, QueryStage_ProtocolInfo
Detail, Node001, Queuing (Query) Get Node Protocol Info (Node=1): 0x01,
0x04, 0x00, 0x41, 0x01, 0xbb
Detail, Node001, Queuing (Query) Query Stage Complete (ProtocolInfo)
Info, Node001, Initilizing Node. New Node: false (false)
Info, contrlr,     Node 009 - New
Detail, Node009, AdvanceQueries queryPending=0 queryRetries=0
queryStage=None live=1
Detail, Node009, QueryStage_ProtocolInfo
Detail, Node009, Queuing (Query) Get Node Protocol Info (Node=9): 0x01,
0x04, 0x00, 0x41, 0x09, 0xb3
Detail, Node009, Queuing (Query) Query Stage Complete (ProtocolInfo)
Info, Node009, Initilizing Node. New Node: false (false)
```

```
Info, contrlr,      Node 010 - New
Detail, Node010, AdvanceQueries queryPending=0 queryRetries=0
queryStage=None live=1
Detail, Node010, QueryStage_ProtocolInfo
Detail, Node010, Queuing (Query) Get Node Protocol Info (Node=10): 0x01,
0x04, 0x00, 0x41, 0x0a, 0xb0
Detail, Node010, Queuing (Query) Query Stage Complete (ProtocolInfo)
```

Then, the system starts adding the new discovered nodes. The reader is first, as follows:

```
-------------------
[DriverReady]:

homeId: 0xe4056d54
nodeId: 1
-------------------


2014-04-24 13:56:31.961 Detail, Node001, Notification: NodeNew


-------------------
[NodeNew]:

homeId: 0xe4056d54
nodeId: 1
-------------------


2014-04-24 13:56:31.963 Detail, Node001, Notification: NodeAdded


-------------------
[NodeAdded]:

homeId: 0xe4056d54
nodeId: 1
-------------------
```

Then, there is the wall plug, as follows:

```
2014-04-24 13:56:31.966 Detail, Node009, Notification: NodeNew


--------------------
[NodeNew]:


homeId: 0xe4056d54
nodeId: 9
--------------------


2014-04-24 13:56:31.967 Detail, Node009, Notification: NodeAdded


--------------------
[NodeAdded]:


homeId: 0xe4056d54
nodeId: 9
--------------------
```

And, in the end, there is the multisensor, as follows:

```
2014-04-24 13:56:31.969 Detail, Node010, Notification: NodeNew

-------------------- [NodeNew]:

homeId: 0xe4056d54
nodeId: 10
--------------------


2014-04-24 13:56:31.972 Detail, Node010, Notification: NodeAdded


--------------------
[NodeAdded]:


homeId: 0xe4056d54
nodeId: 10
--------------------
```

After the probing stage, the system asks for device information, and a lot of it is returned! You can see current values, labels, units of measurement, read-only statuses, and so on:

```
--------------------
[NodeProtocolInfo]:


homeId: 0xe4056d54

nodeId: 1
--------------------


2014-04-24 13:56:32.015 Detail, Node001, Notification: ValueAdded


--------------------
[ValueAdded]:


homeId: 0xe4056d54

nodeId: 1

valueID: 72057594055229441

Value: None

Label: None

Units: None

ReadOnly: False
--------------------


--------------------
[NodeProtocolInfo]:


homeId: 0xe4056d54

nodeId: 9
--------------------


2014-04-24 13:56:32.094 Detail, Node009, Notification: ValueAdded


--------------------
[ValueAdded]:
```

```
homeId: 0xe4056d54
nodeId: 9
valueID: 72057594193723392
Value: False
Label: Switch
Units:
ReadOnly: False
-------------------
...
-------------------
[NodeProtocolInfo]:

homeId: 0xe4056d54
nodeId: 10
-------------------

2014-04-24 13:56:32.150 Detail, Node010, Notification: ValueAdded

-------------------
[ValueAdded]:

homeId: 0xe4056d54
nodeId: 10
valueID: 72057594210680832
Value: False
Label: Sensor
Units:
ReadOnly: True
-------------------
```

As you can see in these examples, the protocol is really powerful and quite complex too! So, to have a better model of our new Z-Wave network, we can use another tool, as follows:

```
root@beaglebone:~/python-openzwave-0.3.0b5/examples# ./api_demo.py
--log=Info --device=/dev/ttyACM0
```

Again, we get tons of messages, but this time, near the end, we can read the following output:

```
Try to autodetect nodes on the network
------------------------------------------------------------
Nodes in network : 3
------------------------------------------------------------
Retrieve switches on the network
------------------------------------------------------------
node/name/index/instance : 9//0/1
   label/help : Switch/
   id on the network : e4056d54.9.25.1.0
   state: False
------------------------------------------------------------
Retrieve dimmers on the network
------------------------------------------------------------
------------------------------------------------------------
Retrieve sensors on the network
------------------------------------------------------------
node/name/index/instance : 10//0/1
   label/help : Sensor/
   id on the network : e4056d54.10.30.1.0
   value: True
node/name/index/instance : 10//1/1
   label/help : Temperature/
   id on the network : e4056d54.10.31.1.1
   value: 0.0 F
node/name/index/instance : 10//3/1
   label/help : Luminance/
   id on the network : e4056d54.10.31.1.3
   value: 675.0 lux
node/name/index/instance : 10//5/1
   label/help : Relative Humidity/
   id on the network : e4056d54.10.31.1.5
   value: 48.0 %
node/name/index/instance : 9//32/1
   label/help : Exporting/
```

```
    id on the network : e4056d54.9.32.1.32

    value: False

node/name/index/instance : 9//4/1

    label/help : Power/

    id on the network : e4056d54.9.31.1.4

    value: 0.0 W

node/name/index/instance : 9//0/1

    label/help : Energy/

    id on the network : e4056d54.9.32.1.0

    value: 0.0 kWh

node/name/index/instance : 9//8/1

    label/help : Power/

    id on the network : e4056d54.9.32.1.8

    value: 0.0 W
-------------------------------------------------------------
Retrieve switches all compatibles devices on the network
-------------------------------------------------------------
node/name/index/instance : 9//0/1

    label/help : Switch All/

    id on the network : e4056d54.9.27.1.0

    value / items: Disabled / set([u'Disabled', u'On and Off Enabled',
u'On Enabled', u'Off Enabled'])

    state: False
-------------------------------------------------------------
Retrieve protection compatibles devices on the network
-------------------------------------------------------------
Retrieve battery compatibles devices on the network
-------------------------------------------------------------
node/name/index/instance : 10//0/1

    label/help : Battery Level/

    id on the network : e4056d54.10.80.1.0

    value : 100
-------------------------------------------------------------
Retrieve power level compatibles devices on the network
```

In this output, it's easier to find all the relative information about our slaves; that is, the wall plug at node 9 (that can work as a normal switch and can return some energy consumption information) and the multisensor at node 10 (that can return the temperature, humidity, and environment luminance and motion activity).

Ok, the Python support is now fully functional! So, let's go to the next section to see how to write the code for our Z-Wave prototype!

The Z-Wave manager

After installing the Python binding to manage Z-Wave devices, we have to write our own code to implement the prototype's software.

As already stated, we need to implement a controller that can record the incoming messages from the sensors, can send some commands to the actuators, and, at the same time, can interact with the user. While the former part is Z-Wave-related, the latter can be implemented by using a web interface created in Python plus some extra HTML/JavaScript and CSS files.

Let's now take a look at the Python code. The whole code is quite long, so I'm going to show the relevant parts only, but you can get the complete code in the `chapter_11/zwmanager.py` file in the book's example code repository.

At the very beginning, we have to declare the code to import:

```
from __future__ import print_function
import os
import sys
import getopt
import string
import syslog
import resource
import time

from openzwave.node import ZWaveNode
from openzwave.value import ZWaveValue
from openzwave.scene import ZWaveScene
from openzwave.controller import ZWaveController
from openzwave.network import ZWaveNetwork
```

```
from openzwave.option import ZWaveOption
from louie import dispatcher, All

from BaseHTTPServer import BaseHTTPRequestHandler, HTTPServer
import json
import cgi
```

As you can see, we need several inclusions from the openzwave package, while BaseHTTPServer, json, and cgi are used to manage the web interface.

Then, some default settings follow:

```
NAME = os.path.basename(sys.argv[0])
debug = False
logstderr = False
log = "Info"
timeout_s = 20
port = 8080

# Default system status
values = {
    "switch" :   "off",
    "power"  :    0.0,
    "temp"   :     0,
    "hum"    :     0,
    "lum"    :     0,
    "bat_lvl":     0,
    "sensor" :   "no",
}
```

Here, the most important thing is the values variable, where we're going to store all devices' statuses.

Then, the Z-Wave-related functions are defined, as follows:

```
def louie_value(network, node, value):
    # Record all new status changing
    if (value.label == "Switch"):
        values["switch"] = "on" if value.data else "off"
    elif (value.label == "Power"):
        values["power"] = value.data
    elif (value.label == "Temperature"):
        values["temp"] = value.data
    elif (value.label == "Relative Humidity"):
        values["hum"] = value.data
```

```
    elif (value.label == "Luminance"):
        values["lum"] = value.data
    elif (value.label == "Battery Level"):
        values["bat_lvl"] = value.data
    elif (value.label == "Sensor"):
        values["sensor"] = "yes" if value.data else "no"
    dbg("dev=%s(%d) name=%s data=%d" % \
        (node.product_name, node.node_id, value.label, value.data))

def louie_network_started(network):
    dbg("network is started: homeid %0.8x" % network.home_id)

def louie_network_resetted(network):
    dbg("network is resetted")

def louie_network_ready(network):
    dbg("network is now ready")
    dispatcher.connect(louie_value, ZWaveNetwork.SIGNAL_VALUE)
```

The relevant functions are the louie_network_ready() function, which installs
a new dispatcher when the Z-Wave network is ready, and the louie_value()
function, which reads all device notifications and stores them in the values variable.

 Note that this code is very far from perfect since we suppose that only one
multisensor device and one wall plug are present at a time! If you wish to
manage more devices, you have to completely review these functions.

Then, the HTTP-related functions follow:

```
class myHandler(BaseHTTPRequestHandler):
    # Disable standard logging messages
    def log_message(self, format, *args):
        return

    # Handler for the GET requests
    def do_GET(self):
        if self.path == "/":
            self.path = "/house.html"
        elif self.path == "/get":
            #dbg("serving %s..." % self.path)

            # Return the current status in JSON format
            self.send_response(200)
```

```python
        self.send_header('Content-type', 'application/json')
        self.end_headers()
        self.request.sendall(json.dumps(values))

        return

    # Otherwise try serving a file
    try:
        # Open the file and send it
        f = open(os.curdir + os.sep + self.path)
        self.send_response(200)
        self.send_header('Content-type', 'text/html')
        self.end_headers()
        self.wfile.write(f.read())
        f.close()
        dbg("file %s served" % self.path)

    except IOError:
        self.send_error(404, 'File Not Found: %s' % self.path)
        dbg("file %s not found!" % self.path)

    return

# Handler for the POST requests
def do_POST(self):
    if self.path == "/set":
        # Parse the data posted
        dbg("managing %s..." % self.path)
        data = cgi.FieldStorage(fp = self.rfile,
            headers = self.headers,
            environ = {'REQUEST_METHOD':'POST',
            'CONTENT_TYPE':self.headers['Content-Type'],})

        self.send_response(200)
        self.end_headers()
        dbg("got label=%s" % data["do"].value)

        # Set the device according to user input
        if data["do"].value == "switch":
        network.nodes[sw_node].set_switch(sw_val,
            False if values["switch"] == "on" else True)

        return
```

```
# Otherwise return error
self.send_error(404, 'File Not Found: %s' % self.path)
dbg("file %s not found!" % self.path)

return
```

The preceding code implements a web server where we just need to manage the GET and POST HTTP requests to do the job. The GET requests are managed by the do_GET() method that simply tries to serve a normal file to the client apart from when the special path /get is used in the URL. In this special case, the server returns the content of the values variable in a JSON format.

As an opposite function, when the POST request is received, it is passed to the do_PUT() method that, in turn, returns an error code if the special path /set is not used; if so, the system parses the data posted by client and then switches the wall plug status.

 Note that we suppose again that only one wall plug is present! So, you have to rewrite the code if you wish to manage more than one wall plug device.

Now, we have to show how the system is set up. After some sanity checks to the command line, we start setting up the Z-Wave network:

```
# Define some manager options and create a network object
options = ZWaveOption(device, config_path = "./openzwave/config",
    user_path = ".", cmd_line = "")
options.set_log_file(NAME + ".log")
options.set_append_log_file(False)
#options.set_console_output(True)
options.set_console_output(False)
options.set_save_log_level(log)
options.set_logging(True)
options.lock()
network = ZWaveNetwork(options, log = None)

# Add the basic callbacks
dispatcher.connect(louie_network_started,
    ZWaveNetwork.SIGNAL_NETWORK_STARTED)
dispatcher.connect(louie_network_resetted,
    ZWaveNetwork.SIGNAL_NETWORK_RESETTED)
dispatcher.connect(louie_network_ready,
    ZWaveNetwork.SIGNAL_NETWORK_READY)
dbg("callbacks installed")
```

```
info("Starting...")

# Waiting for driver to start
for i in range(0, timeout_s):
    if network.state >= network.STATE_STARTED:
        break
    else:
        sys.stdout.flush()
        time.sleep(1.0)
if network.state < network.STATE_STARTED:
    err("Can't initialize driver! Look at the logs file")
    sys.exit(1)

info("use openzwave library   = %s"
    % network.controller.ozw_library_version)
info("use python library      = %s"
    % network.controller.python_library_version)
info("use ZWave library       = %s"
    % network.controller.library_description)
info("network home id         = %s"
    % network.home_id_str)
info("controller node id      = %s"
    % network.controller.node.node_id)
info("controller node version = %s"
    % (network.controller.node.version))
```

The `ZwaveOption()` function is used to set up the network's options, and then the `ZwaveNetwork()` function actually does the job according to the selected options. Then, we set up the callbacks to be called each time a Z-Wave signal arrives, and we have to manage it by using the `dispatcher.connect()` method.

Ok, everything is in place now and we have just to wait for the Z-Wave driver to start. When done, we print some network information. Now, the next step is to wait for the network to be up and running, so we can pass to detect the wall plug device and store its relevant node's information with the following code:

```
# Waiting for network is ready
time_started = 0
for i in range(0, timeout_s):
    if network.state >= network.STATE_READY:
        break
    else:
        time_started += 1
        sys.stdout.flush()
        time.sleep(1.0)
```

```
dbg("detecting the switch node...")
for node in network.nodes:
    for val in network.nodes[node].get_switches():
        data = network.nodes[node].values[val].data
        values["switch"] = "on" if data else "off"
        sw_node = node
        sw_val = val
        dbg(" - device %s(%s) is %s" % \
            (network.nodes[node].values[val].label,
                node,
                values["switch"]))

        # We can manage just one switch!
        break

info("Press CTRL+C to stop")
```

The get_switches() method is used to get all the nodes that can act as a switch, so we use it to detect our wall plug and then to store its information into the sw_node and sw_val variables in order to be used later in the do_POST() method to turn on/off the switch based on the user request.

 Here, it is really clear that the code is written for only one wall plug at a time in the network!

Now, we only have to define the web server to finish the job, and we can do it with the following code:

```
# Create a web server and define the handler to manage the incoming
requests
try:
    server = HTTPServer(('', port), myHandler)
    info("Started HTTP server on port %d" % port)

    # Wait forever for incoming HTTP requests
    server.serve_forever()

except KeyboardInterrupt:
    info("CTRL+C received, shutting down...")
    server.socket.close()
    network.stop()

info("Done.")
```

The main function is the `HTTPServer()` function that starts the internal web server listening at the `8080` port.

Now, to complete the software presentation, I need to show how the `house.html` file works. This is the file that is served by the web server each time a new client gets connected to it.

 Again, as before, I'm going to show the relevant parts only, but you can get the complete code in the `chapter_11/house.html` file in the book's example code repository.

In the head part, I define the CSS file name and the JavaScript code to use:

```html
<head>
    <link href="house.css" rel="stylesheet" type="text/css">

    <script src="/jquery-1.9.1.js"></script>

    <script>
        var polldata = function() {
            $.getJSON('/get', function(data) {
                $.each(data, function(key, val) {
                    var e = document.getElementById(key);

                    if (e != null) {
                        if (e.type == "text")
                            e.value = val;
                    else
                        e.textContent = val;
                    }
                });
            });
        };

        setInterval(polldata, 1000);
    </script>

    <script>
        $(function() {
            $('button[class="do-button"]').click(function() {
                var id = $(this).attr("id");

                $.ajax({
                    url: "/set",
```

```
                        type: "POST",
                        data: "do=" + id,
                        success: function() {
                            console.log('do POST success');
                        },
                        error: function() {
                            console.log('do POST error');
                        }
                    });
                });
            });
        </script>
    </head>
```

In this code, we used the same technique used in *Chapter 7, Facebook Plant Monitor*, where I installed a polling function that executes several GET requests on the server, one per second, to update the displayed data returned in the JSON format. Also, each time a button is pressed, we do a POST request to the server, passing the button ID to it to manage.

In the body part of the house.html file, we define the table to show our data in a nice manner:

```
<body>
 <h1>Home monitor status</h1>

 <h2>Internal variables</h2>

 <table class="status">
  <tr class="d0">
    <td>Switch</td>
    <td><b id="switch">off</b></td>
    <td><button id="switch" class="do-button">switch</button></td>
  </tr>
  <tr class="d0">
    <td>Power[KW]</td>
    <td><b id="power">0</b></td>
    <td></td>
  </tr>
  <tr class="d1">
    <td>Temperature[C]</td>
    <td><b id="temp">0</b></td>
    <td></td>
  </tr>
  <tr class="d1">
```

```
      <td>Relative Humidity[%]</td>
      <td><b id="hum">0</b></td>
      <td></td>
   </tr>
   <tr class="d1">
      <td>Luminance[lux]</td>
      <td><b id="lum">0</b></td>
      <td></td>
   </tr>
   <tr class="d1">
      <td>Battery Level[%]</td>
      <td><b id="bat_lvl">0</b></td>
      <td></td>
   </tr>
   <tr class="d1">
      <td>Motion</td>
      <td><b id="sensor">no</b></td>
      <td></td>
   </tr>
  </table>
</body>
```

Regarding the CSS file, there is nothing important to say (it's just a CSS file!), while the jquery-1.9.1.js file is the one already used in the *The final test*, section in *Chapter 7, Facebook Plant Monitor*; so, just refer to that section in order to know how to get and install it.

The final test

Now, to test the prototype, I connected the wall plug with my printer (the power load) and powered the multi-sensor with a USB port of my PC (just to avoid using the batteries). Then, I started the zwmanager.py program as follows:

```
root@beaglebone:~# ./zwmanager.py -d -l /dev/ttyACM0
zwmanager.py[2732]: callbacks installed
zwmanager.py[2732]: Starting...
zwmanager.py[2732]: network is started: homeid e4056d54
zwmanager.py[2732]: use openzwave library   = 1.3.482
zwmanager.py[2732]: use python library      = 0.3.0b5
zwmanager.py[2732]: use ZWave library       = Static Controller version
Z-Wave 3.79
zwmanager.py[2732]: network home id         = 0xe4056d54
zwmanager.py[2732]: controller node id      = 1
zwmanager.py[2732]: controller node version = 4
```

```
zwmanager.py[2732]: network is now ready
zwmanager.py[2732]: detecting the switch node...
zwmanager.py[2732]:  - device Switch(9) is off
zwmanager.py[2732]: Press CTRL+C to stop
zwmanager.py[2732]: Started HTTP server on port 8080
```

Next, I connected my browser to the `192.168.7.2:8080` URL, but before doing it, I waited for a while, looking at some messages from the sensors:

```
zwmanager.py[4915]: dev=Multi Sensor(10) name=Battery Level data=100
zwmanager.py[4915]: dev=Multi Sensor(10) name=Battery Level data=100
zwmanager.py[4915]: dev=Multi Sensor(10) name=Luminance data=51
zwmanager.py[4915]: dev=Multi Sensor(10) name=Luminance data=51
zwmanager.py[4915]: dev=Multi Sensor(10) name=Relative Humidity data=48
zwmanager.py[4915]: dev=Multi Sensor(10) name=Relative Humidity data=48
zwmanager.py[4915]: dev=Multi Sensor(10) name=Temperature data=20
zwmanager.py[4915]: dev=Multi Sensor(10) name=Temperature data=20
```

Then, when I started the browser, I got the following messages:

```
zwmanager.py[4937]: file /house.html served
zwmanager.py[4937]: file /house.css served
zwmanager.py[4937]: file /jquery-1.9.1.js served
```

As expected, the main HTML file, the CSS file, and the JavaScript file are served to the client that showed me something, as shown in the following screenshot:

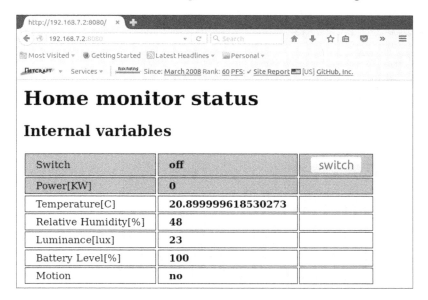

Now, I can try turning on the printer connected to the wall plug by pressing the **switch** button. I got the following messages from the zwmanager.py program:

```
zwmanager.py[5002]: managing /set...
zwmanager.py[5002]: got label=switch
zwmanager.py[5002]: dev=FGWPE Wall Plug(9) name=Switch data=1
zwmanager.py[5002]: dev=FGWPE Wall Plug(9) name=Power data=0
zwmanager.py[5002]: dev=FGWPE Wall Plug(9) name=Power data=21
zwmanager.py[5002]: dev=FGWPE Wall Plug(9) name=Power data=30
zwmanager.py[5002]: dev=FGWPE Wall Plug(9) name=Power data=36
zwmanager.py[5002]: dev=FGWPE Wall Plug(9) name=Power data=13
```

Meanwhile, the web panel changed as follows:

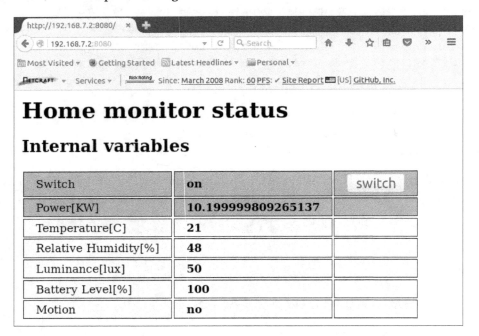

Summary

In this chapter, we discovered how to implement a basic home management system with a web interface that controls two Z-Wave devices to monitor some environment data and control a wall plug.

The presented code, even if a bit complex, can be easily extended to support more Z-Wave devices to manage a really complex network.

Index

Thank you for buying
BeagleBone Home Automation Blueprints

About Packt Publishing

Packt, pronounced 'packed', published its first book, *Mastering phpMyAdmin for Effective MySQL Management*, in April 2004, and subsequently continued to specialize in publishing highly focused books on specific technologies and solutions.

Our books and publications share the experiences of your fellow IT professionals in adapting and customizing today's systems, applications, and frameworks. Our solution-based books give you the knowledge and power to customize the software and technologies you're using to get the job done. Packt books are more specific and less general than the IT books you have seen in the past. Our unique business model allows us to bring you more focused information, giving you more of what you need to know, and less of what you don't.

Packt is a modern yet unique publishing company that focuses on producing quality, cutting-edge books for communities of developers, administrators, and newbies alike. For more information, please visit our website at www.packtpub.com.

About Packt Open Source

In 2010, Packt launched two new brands, Packt Open Source and Packt Enterprise, in order to continue its focus on specialization. This book is part of the Packt Open Source brand, home to books published on software built around open source licenses, and offering information to anybody from advanced developers to budding web designers. The Open Source brand also runs Packt's Open Source Royalty Scheme, by which Packt gives a royalty to each open source project about whose software a book is sold.

Writing for Packt

We welcome all inquiries from people who are interested in authoring. Book proposals should be sent to author@packtpub.com. If your book idea is still at an early stage and you would like to discuss it first before writing a formal book proposal, then please contact us; one of our commissioning editors will get in touch with you.

We're not just looking for published authors; if you have strong technical skills but no writing experience, our experienced editors can help you develop a writing career, or simply get some additional reward for your expertise.

Building Networks and Servers Using BeagleBone

ISBN: 978-1-78439-020-4   Paperback: 110 pages

Set up and configure a local area network and file server by building your own home-based multimedia server

1. Build a DLNA (Digital Living Network Alliance) compatible multimedia server to create your own video theatre and music jukebox.

2. Keep your data safe by setting up a RAID which is used for avoiding data redundancy and to performance improvement.

3. A fast-paced guide that will show you how to set up and configure your own network and file server, with practical steps and illustrations.

Raspberry Pi Home Automation Blueprints

ISBN: 978-1-78328-387-3   Paperback: 360 pages

Create 12 awesome home automation projects using your Raspberry Pi

1. Automate your home devices and appliances with your Raspberry Pi.

2. Use a webcam and the raspicam to view a live stream from a computer or smartphone.

3. Make inexpensive, battery powered remote sensors to detect temperatures, trigger actions, or display charts.

Please check **www.PacktPub.com** for information on our titles

Using Yocto Project with BeagleBone Black

Unleash the power of the BeagleBone Black embedded platform with Yocto Project

Foreword by Khula Azmi, Engineering Manager QA at Mentor Graphics

H M Irfan Sadiq

Using Yocto Project with BeagleBone Black

ISBN: 978-1-78528-973-6   Paperback: 144 pages

Unleash the power of the BeagleBone Black embedded platform with Yocto Project

1. Build real world embedded system projects using the impressive combination of Yocto Project and Beaglebone Black.

2. Learn how to effectively add multimedia to your board and save time by exploiting layers from the existing ones.

3. A step-by-step, comprehensive guide for embedded system development with hands-on examples.

Mastering BeagleBone Robotics

Master the power of the BeagleBone Black to maximize your robot-building skills and create awesome projects

Richard Grimmett

Mastering BeagleBone Robotics

ISBN: 978-1-78398-890-7   Paperback: 234 pages

Master the power of the BeagleBone Black to maximize your robot-building skills and create awesome projects

1. Create complex robots to explore land, sea, and the skies.

2. Control your robots through a wireless interface, or make them autonomous and self-directed.

3. This is a step-by-step guide to advancing your robotics skills through the power of the BeagleBone.

Please check **www.PacktPub.com** for information on our titles

www.ingramcontent.com/pod-product-compliance
Lightning Source LLC
LaVergne TN
LVHW081330050326
832903LV00024B/1107

9781783986026